Boukar Michel

Modeling CO$_2$ emissions

Boukar Michel

Modeling CO$_2$ emissions

due to the consumption of petroleum products in Chadian road transport Preface by Mr. TAMBA Jean Gaston

ScienciaScripts

Imprint

Any brand names and product names mentioned in this book are subject to trademark, brand or patent protection and are trademarks or registered trademarks of their respective holders. The use of brand names, product names, common names, trade names, product descriptions etc. even without a particular marking in this work is in no way to be construed to mean that such names may be regarded as unrestricted in respect of trademark and brand protection legislation and could thus be used by anyone.

Cover image: www.ingimage.com

This book is a translation from the original published under ISBN 978-620-6-72127-7.

Publisher:
Sciencia Scripts
is a trademark of
Dodo Books Indian Ocean Ltd. and OmniScriptum S.R.L publishing group

120 High Road, East Finchley, London, N2 9ED, United Kingdom
Str. Armeneasca 28/1, office 1, Chisinau MD-2012, Republic of Moldova, Europe
Printed at: see last page
ISBN: 978-620-8-10746-8

MODELING OF CO_2 EMISSIONS DUE TO PETROLEUM PRODUCT CONSUMPTION IN CHADIAN ROAD TRANSPORT

BOUKAR MICHEL

Foreword by **TAMBA Jean Gaston**
Full Professor - University of Douala

PREFACE

The thesis by M. Boukar Michel's thesis on "Modelling CO_2 emissions due to petroleum product consumption in Chadian road transport" is a remarkable study whose publication is to be welcomed as an important contribution to highlighting the short- and long-term dynamics of petroleum product consumption (super and diesel) on CO_2 emissions in Chad, taking into account petroleum product prices (super and diesel), the vehicle fleet and urbanization in Chad; to predict CO_2 emissions in Chad on the basis of petroleum product consumption, petroleum product prices, vehicle fleet and urbanization, and ultimately to help Chadian policy-makers formulate policies and corresponding measures to ensure effective control of CO_2 emissions in the country.

At the end of this study, Mr. Boukar Michel was able to achieve the four objectives of his study. He was able to highlight the long- and short-term dynamics of all the variables under study, and the existing causal analysis between the variables used in the study was provided. Mr. Boukar Michel was able to predict future carbon dioxide emissions, and public recommendations were issued to the Chadian government.

With this work, praised by a prestigious jury at the University of Douala and now published by European university students, the author is well equipped to take up the challenge of a promising academic career.

TAMBA Jean Gaston

Full Professor of Energetics

Douala University

DEDICATION

A

MY WONDERFUL FATHER BOUKAR VOUSOUMA MAURICE DECEASED

NOVEMBER 15, 2023, AND HAS NOT BEEN ABLE TO SEE MY WORK.

ACKNOWLEDGEMENTS

The completion of this thesis is the fruit of many combined efforts. I would like to express my sincere thanks to all those who have contributed in any way to the realization of this scientific work. In particular :

- To Professor ONDOA Magloire, Rector of the University of Douala, for providing me with a stimulating research environment conducive to intellectual growth.

- To Professor KWATO NJOCK, for his leadership and vision, which have enabled the Ecole Doctorale des Sciences Fondamentales et Appliquées (EDOSFA) to become a center of research excellence.

- To the Coordinator of the Unité de Formation Doctorale des Sciences Appliquées (UFD-SCA), Professor TOMEDI EYANGO, for her involvement in the life of the UFD and her commitment to the success of doctoral students.

- To my thesis supervisor, and also Coordinator of the Transport and Applied Logistics Laboratory (LTLA), Professor TAMBA Jean Gaston, who has done me the honour of agreeing to supervise this work. I would like to express my deepest gratitude.

- To the Chairman of the Jury, who has done me the honor of presiding over this presentation.

- To all the rapporteurs and members of the jury, for their valuable comments and contributions, and who will take the trouble to examine this work with the aim of contributing to its improvement. I am thinking in particular of : Pr. FOPAH-LELE Armand, Pr. KOUMI NGOH Simon and Pr. MOUZOUG PEMI Marcelin, who did me the honor of acting as rapporteurs for my thesis, and who took the time to listen to me and discuss it with me. Their comments enabled me to see my work from a different angle. For all this, I thank them.

- To Dr. SAPNKEN Flavian Emmanuel, for his constant support and the precious time he has given me since the beginning of this thesis.

- To all my teachers at LTLA, for their unceasing support and their many tips which enabled me to complete this work.

- I would also like to thank Mr. DANWE Raidandi, Coordinator of the Doctoral Training Unit at the Ecole National Supérieur Polytechnique de Maroua and Full Professor at the University of Maroua, who welcomed me to his laboratory for two years. Thanks to him, I was able to happily combine theoretical and applied research during my research.

- I would like to thank Prof. MAHAMAT Barka, who taught me optics and thermodynamics 30 years ago at the Faculty of Exact and Applied Sciences and, today, for agreeing to take part in my thesis jury and for his scientific participation and the time he has devoted to my research.

- I would also like to take this opportunity to thank His Excellency, General MAHAMAT Idriss Deby ITNO, President of the Transition, President of the Republic of Chad, Head of State, for having contributed in various ways to the success of this thesis.

- I would also like to thank Dr. DAVID Houndeigar, former Secretary General of the Presidency of the Republic of Chad, who inspired the title of this thesis.

- I can't forget Madame FATIMA Goukouni, Chad's Minister of Transport, Civil Aviation and Metrology, for her invaluable help in collecting data within her Ministry, not forgetting her entire team at the Ministry for their invaluable assistance.

- I would also like to say a big THANK YOU to my wife Mariam Sanah Issa, for her availability, patience and tolerance, which has enabled me to complete this thesis today.

- I would like to thank Mr. TOM ERDIMI, Minister of State, Minister of Higher Education of Chad, for having contributed to the knowledge of science, some 33 years ago when we entered the Faculty of Exact and Applied Sciences, I remember your first lectures in general mechanics in the Amphi Thibeau. Neither I nor anyone else could have imagined that one day I would be able to produce scientific books and defend my thesis today in your presence.

- You have contributed to the training of our generation, including those who are no longer with us: Dr TAGUI Kelbeye, Dr IBNI Oumar Mahamat saleh, Dr TCHADANAYE New Mahamat. This list is not exhaustive. May their souls rest in peace.

- My thanks also go to Dr ALI Abdramane Haggar, Prof. MAHAMAT Daoussa Haggar, President of the University of N'djamena, Prof. MAHMOUD Youssouf Khayal, former Minister of Health of Chad, and Head of the Doctoral Training Program in Engineering Science, and Prof. MAMOUT YAYA, Head of the Doctoral Training Program in Chemistry, who have done everything to help me, who have supported me and above all supported me in everything I have undertaken.

- I can't forget my classmate Pr BASSA Bruno, Antipas, Dr Mahamat Ahmat Alhabo, currently Minister of State, Secretary General to the Presidency of the Republic of Chad, and Pr Banbo Antipas, Rector of the Académie du Sud, who encouraged me to complete this thesis.

- A special thanks to my friend Ahmat Khazali ACYL, with whom I did this thesis. Find here, my dear, the words of friendship and affection of your friend.

- Special thanks to my childhood friends Malachie Ngueffa and Etienne Balamto, with whom I have fond memories.

- To all my classmates, especially Mr EWODO AMOUGOU Marcel Rodrigue and Mr SAMA Jean Marie Stevy, for having spared no effort to bring a particular analysis to this work.

TABLES OF CONTENTS

SUMMARY

Emissions of CO_2 emissions in Chad have been growing steadily over the years, and the safeguarding of the Paris climate agreements in April 2016 and January 2017 is undeniable. It is therefore important to carry out a life-saving study highlighting specific recommendations for controlling emissions. CO_2. Thus, this thesis examines the short- and long-term dynamics of petroleum product consumption (super and diesel), petroleum product prices (super and diesel), the vehicle fleet and urbanization on emissions of CO_2 emissions in Chad; it provides an analysis of the causal links between CO2 emissions, petroleum product consumption, petroleum product prices, the vehicle fleet and urbanization in Chad; and finally, it forecasts future emissions of CO_2 emissions in Chad based on petroleum product consumption, petroleum product prices, vehicle fleet and urbanization. This thesis is based on a solid literature review, and the unit root test was used to examine the stationarity of the series firstly. Secondly, ARDL models were estimated and robustness tests carried out to attest to the validity of the models. Thirdly, bounds tests were applied to confirm the existence of long-term relationships. Fourthly, Granger causality tests in the Toda-Yamamoto sense were applied to establish the various influences existing between the substudy variables. Finally, a hybrid model was used for forecasting: the sequential GMC (1, N) - GA. The results of model 1 show that, in the long term, a 1% increase in supermarket consumption would lead to a 1.03% increase in emissions of CO_2 emissions, and that there is a unidirectional causality from supermarket consumption to CO_2. In Model 2, in the long term, the linear impact of diesel consumption on CO_2 emissions is positive and insignificant. There is bidirectional causality between diesel consumption and emissions of CO_2. Finally, future forecasts based on the GMC (1, N) - GA sequence show that emissions of CO_2 emissions will be 2672.39 kilotonnes by 2030. Following this work, these results could be useful for guiding environmental policies aimed at reducing emissions of CO_2 emissions through mechanisms to reduce petrol vehicle fuel consumption and control diesel consumption using appropriate tools.

Keywords: *Emissions of CO_2; Consumption of petroleum products; Road transport; ARDL; Le Séquentiel GMC (1, N) - GA, Chad.*

GENERAL INTRODUCTION

The transport sector has been essential for the movement of people and the supply of goods and services needed in daily life for years (Ağbulut, 2022; Onat et al., 2014; Tamba et al., 2012). It also has an impact on all aspects of human presence, including production, culture, trade, education and defense (Danish et al., 2018). This sector has seen significant growth thanks to energy consumption, particularly that of petroleum products (Engo, 2019). Indeed, global fossil fuel consumption in the transport sector rose from 88074462 TJ in 2005 to 110471146 TJ in 2019, an increase of 25.43% (IEA, 2021). Moreover, the transport sector is one of the main final consumers of total energy contributing to global emissions. It contributes 29% of total global energy consumption and 65% of global consumption of petroleum products (Solaymani, 2019). As a result, demand for petroleum products in the transport sector has become increasingly important over the years. Given that the economic performance of any nation is measured by its rate of energy consumption, it follows that this demand plays an important role in a nation's economy (Sapnken, 2018; Carbonnier and Grinevald, 2011).

Chad's overall consumption of fossil fuels between 2010 and 2016 was high, resulting in high greenhouse gas emissions. (TCNTCC, 2020). In 2010, gross energy consumption of fossil fuels in Chad, excluding biomass, was estimated at 257.01ktoe, of which 91.55ktoe was diesel consumption, representing 35.62%, and 16.91ktoe was gasoline consumption, representing 6.58%. (TCNTCC, 2020). For 2016, gross energy consumption of fossil fuels in Chad, excluding biomass, was estimated at 662.86ktoe, i.e. 331.87ktoe for diesel consumption, representing 50.07%, and 115ktoe for super gasoline consumption, representing 17.49%. (TCNTCC, 2020). These figures show that diesel consumption in Chad has increased by 14.45%, while super diesel consumption has also risen by 10.91% over the 7-year period from 2010 to 2016. In the road transport sector, Chad's consumption of petroleum products such as super and diesel amounted to 8916m³ in 2007, compared with 130275m³ in 2016, an increase of 121359m³ in 10 years. The consumption of petroleum products in the road sector has increased due to the considerable number of vehicles registered each year and the rapid growth in the population. Indeed, in 2016, Chad had nearly 245601 vehicles of all types over the last ten years, and the population growth rate between 1985 and 2000 was 2.4%, compared with 3.9% between 2000 and 2015. (TCNTCC, 2020).

Greenhouse gas emissions in Chad come from fuel consumption in the transport sector. To this must be added the consumption of fuels for electricity generation, fuels for domestic use and fugitive emissions in the oil sector. (TCNTCC, 2020). Carbon dioxide is the most emitted greenhouse gas in the atmosphere (Engo, 2019). In this respect, carbon emissions can cause numerous health and environmental problems (Dong et al., 2021). From the health point of view, they can cause serious respiratory problems such as breathlessness, headaches and fatigue, and from the environmental point of view, they can cause climate change, acid rain affecting the environment. Near-real-time data show that global emissions of CO_2 emissions rose by 4.8% in 2021 to 34.9 GtCO_2 following record declines in 2020 (Liu et al., 2022). Worldwide, transport as a whole was responsible for 23% of total emissions of CO_2 emissions from fuel combustion, and road transport was responsible for 20% in 2014 (Santos, 2017). Even the work of Solaymani (2019) support the estimates for this sector, which accounts for around 24% of global greenhouse gas emissions. CO_2 emissions, also due to fuel combustion. Generally speaking, the land transport sector is recognized worldwide as one of the biggest emitters of greenhouse gases. (TCNTCC, 2020). Land transport is still largely fuelled by fossil fuels, and is therefore responsible for significant emissions of greenhouse gases, notably carbon dioxide (CO_2), and atmospheric pollutants (Ehrenberger et al., 2021). In fact, emissions of CO_2 emissions were 304.99ktoe in 2000, compared with 1938.59ktoe in 2016; this shows that emissions of CO_2 emissions in Chad increased sixfold between 2000 and 2016. According to the greenhouse gas inventories carried out in five sectors, including the energy sector alongside industrial processes, agriculture, land use, land use change and forestry (LULUCF) and waste, with 2010 as the reference year, Chad continues to be a source of carbon ten years after the inventories carried out as part of the initial national communication and the second national communication. (TCNTCC, 2020).

The United Nations Framework Convention on Climate Change (UNFCCC) having been signed and ratified at the Rio Earth Summit in June 1992 and April 1993, and ratifying the Kyoto Protocol (KP) in August 2009 and the Paris Climate Agreement in April 2016 and January 2017, Chad is consolidating its commitment to work towards safeguarding the global climate, a factor in development, and has set itself the objective of reducing its greenhouse gas emissions, notably emissions of CO_2 (CDN, 2021; TCNTCC, 2020). There is no doubt that policies to reduce carbon emissions will have a significant influence on achieving this vision. However, reducing carbon emissions means lowering energy consumption, which will hamper the country's economic growth. As a result, energy consumption and the problem of reducing carbon emissions in the road transport sector need to be considered together, along with the forecast of carbon emissions due to the consumption of petroleum products in the road transport sector. Our analysis therefore aims to answer the question: what is the impact of petroleum product consumption in the road transport sector on carbon emissions in Chad? CO_2 emissions in Chad? What are the causal links between emissions of CO_2 emissions and the consumption of petroleum products by the Chadian road sector? What are the predicted values of CO_2 emissions based on petroleum product consumption, petroleum product prices, vehicle fleet and urbanization? To address these concerns, this analysis aims to explore the association between CO_2 emissions and petroleum product consumption in Chad's road sector over the 2008-2019 period, incorporating other variables such as petroleum product prices, the vehicle fleet and urbanization. On the other hand, this work aims to estimate future emissions of CO_2 emissions in Chad, taking into account petroleum product consumption, petroleum product prices, the vehicle fleet and urbanization. Specifically, the first step is to examine the short- and long-term effects of an increase in the consumption of petroleum products by the road sector on emissions of CO_2 emissions in Chad, taking into account the price of the products concerned, the vehicle fleet and urbanization. Secondly, we will examine the influences that exist between emissions of CO_2 emissions and the consumption of petroleum products in the road sector, taking into account the price of petroleum products, the vehicle fleet and urbanization in Chad. Finally, we will assess future values of CO_2 emissions in Chad. The objectives of this study are :

i. Present the general situation of greenhouse gas emissions and make an analysis of the road sector in Chad;

ii. Highlight the short- and long-term dynamics of petroleum product consumption (super and diesel), petroleum product prices (super and diesel), vehicle fleet and urbanization on emissions of CO_2 emissions in Chad ;

iii. Provide an analysis of the causal links between emissions of CO_2 emissions, petroleum product consumption, petroleum product prices, the vehicle fleet and urbanization in Chad;

iv. Predict future emissions of CO_2 emissions in Chad based on petroleum product consumption, petroleum product prices, vehicle fleet and urbanization;

v. To help Chadian decision-makers formulate policies and corresponding measures to ensure effective control of the country's greenhouse gas emissions. CO_2 emissions in the country.

This study contributes to the current literature in several ways: (i) To the best of our knowledge, this work represents a new attempt to simultaneously highlight the short- and long-term effects of petroleum product consumption, crude oil prices, the vehicle fleet and urbanization on emissions of CO_2 emissions in Chad; (ii) This study integrates the analysis of causality between emissions of CO_2 emissions, petroleum product consumption, crude oil prices, the vehicle fleet and urbanization in Chad, which has not been covered to date; (iii) This study uses fairly recent data, depending on their availability in Chad; (iv) This study is intended for Chadian political leaders, to help them formulate policies regarding petroleum product consumption in the road transport sector, in order to reduce emissions of CO_2 emissions and ensure sustainable development.

The period from 2008 to 2019 was chosen in view of the official data structure in Chad, which provides this information at this time interval.

The remainder of this thesis is organized as follows: The first chapter presents a brief overview of the environmental aspect in Chad. Next, it carries out an analysis of the Chadian road transport sector; and finally, it establishes an overview of previous work and justifies the parameters used. Highlights of the methodological framework and data sets used in this study are discussed in Chapter 2. The framework describes step-by-step how to establish models for estimating short- and long-term effects, causality inferences in an augmented VAR following the Toda-Yamamoto procedure, and the forecasting method used to determine future emissions of CO_2. This forecasting method is based on the GMC (1, N) - GA sequential model. Finally, Chapter 3 presents and analyzes the empirical results. We conclude this study by discussing policy implications to guide the government of Chad in adopting effective policy measures to reduce emissions of . CO_2.

CHAPTER I
GENERAL

Introduction

This chapter briefly presents the environment in Chad and analyzes the country's road transport sector. This will be followed by a literature review highlighting the body of previous work on the linear and causal relationships of petroleum product emissions in road transport. CO_2 emissions due to the consumption of petroleum products in road transport will be presented, preceded by a set of justifications for the parameters contained in the econometric model used.

I.1A brief overview of the environmental aspect

I.1. 1Greenhouse gas (GHG) emissions worldwide

Global GHG emissions rose from 38,669 to 48,117 megatonnes of carbon dioxide equivalent (Mt CO eq. CO_2), an increase of 23.6%. (World Resources Institute, 2022). This has resulted in a 1.2°C rise in average global surface temperature (World Meteorological Organization, 2021).. The Intergovernmental Panel on Climate Change (IPCC) published scientific evidence of the effects of global warming in its first assessment report in 1990. The United Nations Framework Convention on Climate Change (UNFCCC) was adopted in 1992 with the aim of stabilizing GHG concentrations in the atmosphere to avoid dangerous interference with the climate system. (UN, 1992). The Kyoto Protocol was adopted at COP3 in 1997, with the aim of reducing greenhouse gas emissions. EMISSIONS (UNFCCC, 2009).

At the 18th session of the Conference of the Parties (COP18) in Doha, one of the important outcomes was nationally appropriate mitigation measures. These measures are designed to encourage developing countries to reduce their emissions by 2020. The outcome of COP21 is the Paris Agreement, a new international treaty under the UNFCCC dealing with GHG mitigation, adaptation and financing. The Paris Agreement aims to limit the increase in global surface temperature to less than 2°C compared to the pre-industrial revolution (Guilyardi et al., 2018; Pita et al., 2017). In order to achieve the Paris Agreement's objective, the results of COP24 in Katowice provide the rules for meeting the Paris Agreement's requirements in key areas.

Achieving the Paris Agreement is uncertain. Results indicate that the emissions target limiting global surface temperature increase to below 2°C by 2100 will not be met under current policies (Pita et al., 2020). To limit warming to less than 2 degrees Celsius, GHG emissions must be reduced rapidly and brought to zero by mid-century. In addition, the IPCC special report on 1.5°C suggests energy mitigation measures to promote renewable energies and carbon storage (Pita et al., 2020). Long-term targets to achieve net zero emissions worldwide by 2050 require urgent action and nationally determined contributions (Akeresola and Gayawan, 2021).

I.1.2 Situation of greenhouse gas emissions in Africa

Africa, as a continent, has been the area most affected by the negative effects of global warming (Bewket, 2012). Warmer temperatures and reduced rainfall in Tanzania led to a 49% loss of ice on Mount Kilimanjaro between 1976 and 2002. This has led to the drying up of most of the surrounding rivers (Josephine, 2007). If the increase in GHG emissions is not curbed, forecasts indicate that there will be no ice left on Mount Kilimanjaro, Africa's highest mountain, by 2033. Change (2008) indicated that Tunisia is the country most likely to experience a decrease in annual precipitation by 2030 and a 2.1°C increase in average temperature by 2050. As a result, the risk of water shortages will be high. And, in turn, harm the country's inhabitants.

Africa is the area that contributes the least to total global GHG emissions (Adzawla et al., 2019; Bewket, 2012).. However, anthropogenic GHG emissions are rising steadily in Africa. In East Africa, total GHG emissions rose by 42% between 1990 and 2011. West Africa also recorded a 17% increase in GHG emissions between 1990 and 2014, with Nigeria contributing around 46% of these emissions (Akeresola and Gayawan, 2021). By 2050, projected GHG emissions in sub-Saharan Africa will increase by up to 50%. (Leimbach et al., 2018)and double for Africa as a whole (van der Zwaan et al., 2018).

The IPCC states that most of the observed increase in global average temperatures since the mid-20th century is very likely due to the observed increase in man-made greenhouse gas emissions. According to the United Nations Framework Convention on Climate Change (UNFCCC), the main GHGs are carbon dioxide (CO_2), methane (CH4), nitric oxide (N$_2$ O) and fluorinated gases (per-fluorocarbons (PFCs), hydro-fluorocarbons (HFCs) and sulfur hexafluoride (SF6)).

Africa is pursuing policies that will help minimize its GHG emissions, given that GHGs have a long atmospheric lifetime of several thousand years (Akeresola and Gayawan, 2021). However, efforts to mitigate GHG emissions in developing economies are limited by the commitment of financial and technological resources

(Adzawla et al., 2019). Furthermore, the world's top ten poorest countries in 2020 are from Africa (Akeresola and Gayawan, 2021). Therefore, policy effectiveness will require understanding how these GHG emissions evolve simultaneously in space and time and the impact of the interaction of space and time on these emissions in Africa in order to maximize these minimal resources.

I.1.3 Greenhouse gas emissions in Chad
a) Geographical features

Chad is a Central African country covering a vast area of 1,284,000 km² and extending between 7° and 24° North latitude and between 14° and 24° East longitude (Bedoum et al., 2017).. It is a country bordered to the north by Libya, to the east by Sudan, to the south by the Central African Republic and to the west by Cameroon, Nigeria and Niger. It is therefore bounded by six countries. Fig. 1 shows the geographical map of Chad. Chad's relief is characterized by natural ensembles (Bedoum et al., 2017; UNFCCC, 2001). These include:

- Southern Chad, south of the 10th parallel, corresponding to the high basins of the Chari and Logone rivers, with an average altitude of 400 to 500 m, and mountain ranges culminating at 1163 m ;
- The Logone floodplains between Lai and NDJAMENA (300 to 400 m altitude) and the south-eastern floodplains along the CAR border (400 to 450 m);
- The Guera massif in the center, culminating at 1500 m ;
- The Chari deltas, at altitudes ranging from 300 to 350 m, with ancient deltaic formations of clayey-sandy alluvium;
- Floodplains and dune belts around Lake Chad with altitudes ranging from 280 to 290 m ;

- The Ouaddaï massif to the east (500 to 1000 m);
- And finally, the Tibesti massif in the north, with the Emi Koussi peak reaching an altitude of 3,415 m. This is also the part of the country with the lowest depressions (160 m altitude).

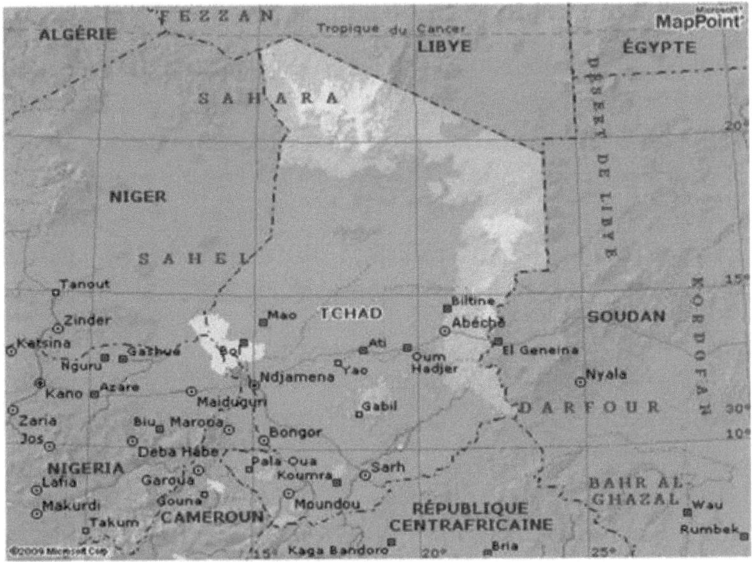

Figure 1 Map of Chad
Source: Bedoum et al (2017)

b) Climate

The Chadian territory is subject to a climate that depends on the rainfall regime, the succession of seasons and the thermal regime. Fig. 2 shows that this climate can be Saharan between 14° and 24° North latitude, Sahelian between 10° and 14° North latitude and Sudanian at latitudes 7°-10° **North** (Bedoum et al., 2017; UNFCCC, 2001).

The Saharan or desert zone in the north covers the northern part of the country and comprises the provinces of Borkou-Tibesti-Ennedi (BET) and the northern parts of the provinces of Kanem and Batha, i.e. 47% of the national territory. With rainfall below 100 mm/year, only oasis agriculture and camel and small ruminant farming are practiced. The season lasts two months in the north (almost nil in the far north). (CDN, 2021; TCNTCC, 2020).

The Sahelian zone in the center of the country covers 43% of the national territory. With rainfall ranging from 100 to 800 mm/year, it presents major contrasts between the arid northern part (Saharo-Sahelian climate with annual rainfall ranging from 100 to 200 mm) and the Sahelo-Sudanian zone in the south, characterized by rainfall ranging from 600 to 800 mm/year. (CDN, 2021; TCNTCC, 2020).

The Sudanian zone in the south of the country, between isohyets 800 and 1,200 mm, represents only 10% of the national territory. However, it is home to almost half of Chad's population. The sub-humid tropical climate, with rainfall in the Sudano-Guinean zone exceeding 1,200 mm, allows for a wide variety of agricultural production and livestock breeding (cattle, goats, sheep, pigs, poultry). (CDN, 2021; TCNTCC, 2020).

The climate of the Republic of Chad is similar to that of West African countries, generated by the displacement of the Intertropical Convergence Zone (ITCZ) with :
- A very hot period from March to June, when temperatures can reach 45°C;
- A rainy season from May to October, with average rainfall between 50 and 1350 mm;
- And a relatively dry, cold period from November to February. Rainfall in Chad is generally characterized by a highly irregular spatial and temporal distribution from one station to another.

Generally speaking, the different climates encountered in Chad are characterized by high temperatures, generally fluctuating between 30 and 40°C in maximum daily values, wide daily temperature ranges, always in excess of 10°C, but often close to 20°C, and precipitation that varies greatly according to latitude. Interannual temperature index trends from 1950 to 2019 show a steady rise in temperature from the early 1980s to the present day. On a global scale, 1990 and 2000 were the hottest years since meteorological records began in Chad. Maximum temperatures rose by an average of 1.1°C across the country. Minimum temperatures are estimated to have risen by 2°C over the period 1951-2010, and maximum temperatures by 1°C, with high values between 2002-2010 (CDN, 2021).

This succession of climates is explained by the successive dominance of trade winds moving dry air masses from the northeast (desert) and monsoon winds moving humid air masses from the southwest (Gulf of Guinea). The meeting of these air masses creates an intertropical convergence front that moves alternately from the north to the south of the country. To the north of the front is a zone of dry trade winds, with clear skies and sandy winds; at the level of the front, the clash of air masses is responsible for tornadoes and occasional rainfall; to the south of the front, the humid air masses are responsible for heavy and frequent rainfall (squall lines). Monsoon rains are generally violent and intense, particularly in tropical and Sahelian zones, with annual 24-hour maxima of 70 mm in Sarh, 58 mm in NDJAMENA and 50 mm in Abéché.

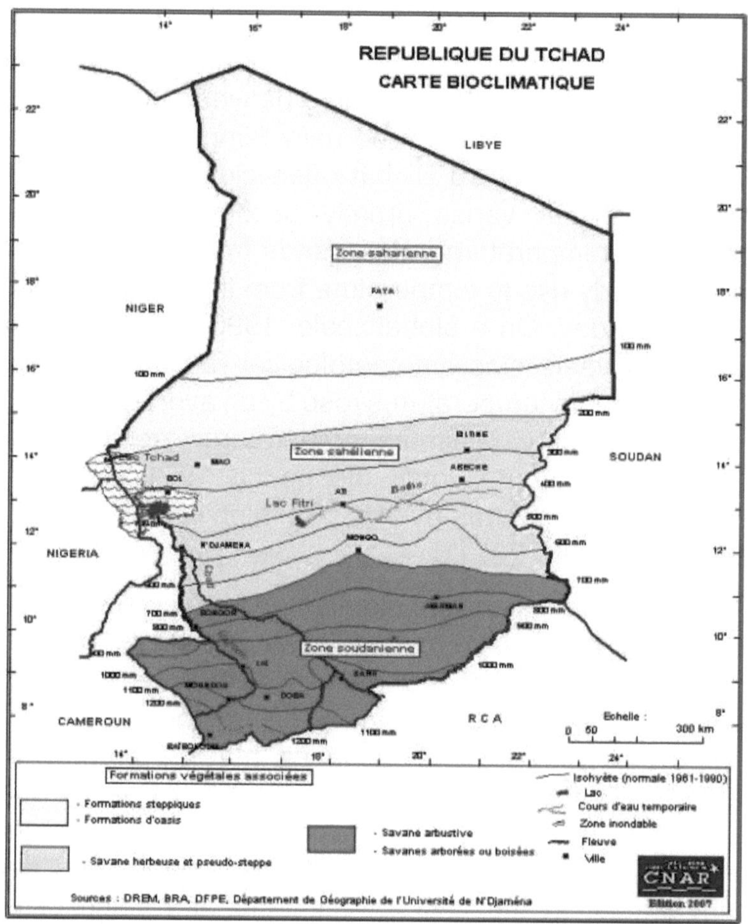

Figure 2 The three main bioclimatic zones in Chad
Source : CDN (2021)

Winds are generally from the north-east to the south-west, with average speeds of between 20 and 40 km/h. As in all of Sahelian Africa, Chad's climate is subject to considerable inter-annual fluctuations. What's more, over the past two decades, Chad has experienced a series of particularly dry years, resulting in a southward shift of isohyets, and therefore of climatic zones as they are conventionally defined.

c) GHG situation in Chad

Several sectors are responsible for greenhouse gas emissions, including energy, agriculture, forestry and waste, from 2010 to 2018. Greenhouse gas emissions rose by 50%, from 49,320 kt of eq. CO_2 in 2010 versus 74,090 kt of eq. CO_2 in 2018 (CDN, 2021). Greenhouse gas emissions come mainly from fuel consumption for electricity generation, fuel consumption for the transport sector, fuel consumption for domestic use and fugitive emissions in the oil sector. Fig. 3 shows that in Chad, emissions of CO_2 emissions are growing year on year. Over the decade from 2000 to 2010, emissions of CO_2 emissions more than doubled from 304.99 ktoe to 695.17 ktoe. In 2016, emissions of CO_2 emissions have almost tripled compared to 2010.

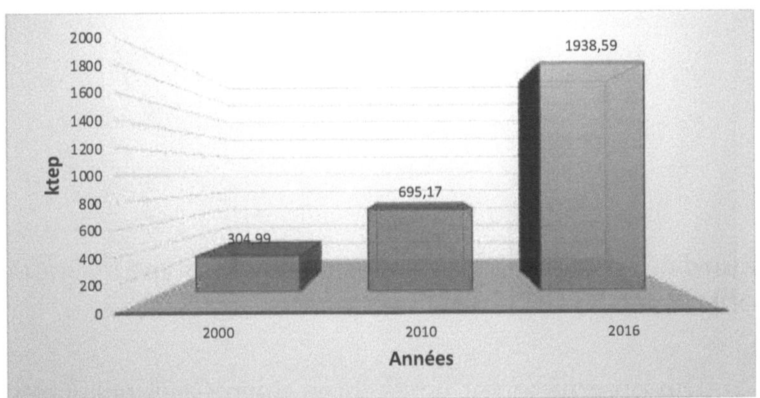

Figure 3GHG emissions trends
Source : TCNTCC (2020)

GHG emissions in 2010 were dominated by fugitive emissions from flared gases at oil sites. These are followed by diesel combustion, used in particular in the electricity and transport sectors. Gasoline and jet fuel combustion come third and fourth in terms of GHG emissions. EMISSIONS (TCNTCC, 2020). In contrast to the reference year, in 2016 emissions of CO_2 emissions are largely dominated by diesel combustion. Emissions from gasoline combustion are in second place, and are mainly emitted by the land transport sector. LPG consumption in the residential sector has grown exponentially, particularly in the capital. Emissions from air transport have also risen by almost 50%. Fugitive emissions, on the other hand, have fallen considerably. Fig. 4 shows emissions of CO_2 emissions by fuel type in GgCO_2.

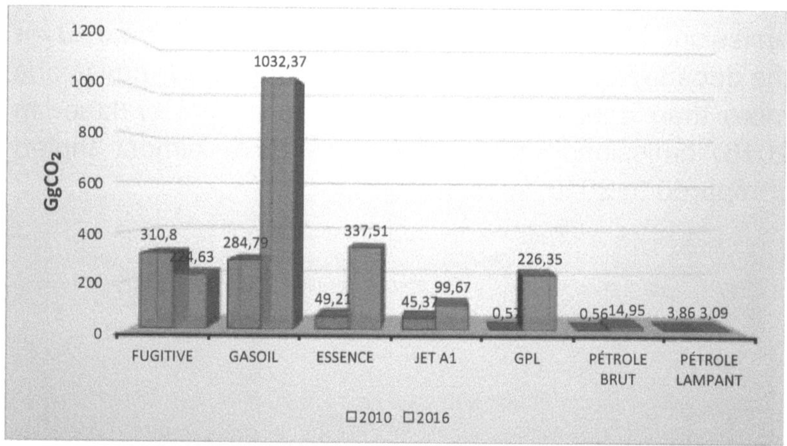

Figure 4 Emissions of CO_2 emissions by fuel type in GgCO_2
Source : TCNTCC (2020)

The energy sector emits three main GHG pollutants: CO CO_2, CH_4 and N_2O. Using the baseline approach, this study estimates only the total emissions of CO_2 emissions attributable to this sector. Table 1 summarizes GHG emissions from 2010 to 2016 in Chad as follows:

Table 1 Gross GHG emissions from 2010 to 2016

POLLUTANTS	2010		2016	
	Emissions in CO_2 eq (Gg)	% of total energy excluding biomass	Emissions CO_2-eq (Gg)	% of total excluding biomass
CO_2	695.17	100%	1938.59	100%
CH_4	-		-	-
N_2O	-		-	-
PRG	695.17		1938.59	

Source : TCNTCC (2020)

GHG emissions from the energy sector for the 2010 reference year are estimated at 695.17 Gg Eq. CO_2. These emissions are significantly lower than those for 2016, which amounted to 1938.59 Gg Eq. CO_2. This rise in emissions is mainly due to electricity generation, oil production activities and the transport sector. Given the almost negligible emissions from CH_4 and N_2O, the CO_2 represents 100% of GHG emissions in the energy sector.

I.2 Analysis of the transport sector

I.2.1 Analysis of road transport

In Chad, the road transport sector is constantly growing, given the large number of vehicles registered each year. In 2016, the number of vehicles registered over the last ten years stood at 24,5601 (all vehicles combined), broken down as follows: 172486 motorcycles; 43942 light diesel vehicles; 19480 light petrol vehicles and 9693 heavy goods vehicles such as trucks, tractors and semi-trailers. (TCNTCC, 2020). It should be noted that, well before 2007, there were obviously registered vehicles, but their numbers have not been included in this estimate. Given their age, these vehicles, if still in circulation, would be the biggest emitters of greenhouse gases. Fig. 5 shows statistics on the fleet of registered vehicles from 2007 to 2016.

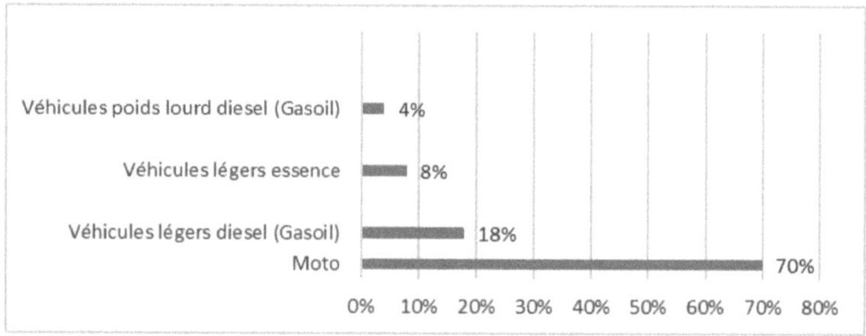

Figure 5:Statistics on the fleet of registered machines from 2007 to 2016
Source : TCNTCC (2020)

In this research, the vehicle fleet is made up of passenger cars, vans, buses, rigids, tractors, semi-trailers, trailers and machines. Fig. 6 shows the evolution of the vehicle fleet in Chad from 2008 to 2019. It can be seen that the number of vehicles in Chad fell considerably between 2008 and 2011. This drastic drop in the number of cars in Chad can be explained by the rise in the price of petroleum products during this period. This rise in the price of petroleum products is based on Chad's 100% imports of petroleum products, mainly from Cameroon and Nigeria. Following the commissioning of the NDJAMENA refinery, the number of cars on the road in Chad increased significantly in 2012-2013. This increase is explained by the halving of petroleum product prices and the considerable drop in petroleum product imports. From 2014 onwards, the number of vehicles on the road in Chad will decline considerably until 2019.

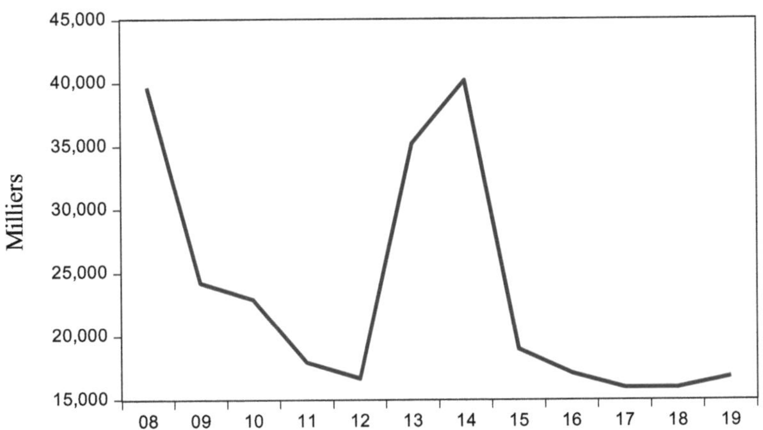

Figure 6 Evolution of the vehicle fleet in Chad from 2008 to 2019

Fuel consumption for land transport is based on a number of simplifying assumptions, taking into account the realities of the country. TCNTCC (2020) informs us that fuel consumption for land transport over the period 2007-2016 is estimated at 30 liters per month for motorcycles on average; 100 liters per month for gasoline-powered light trucks on average; 100 liters per month for diesel-powered light trucks on average; and 150 liters per month for diesel-powered heavy trucks on average. Diesel-powered light commercial vehicles ranked first in terms of fuel consumption over the 2007-2016 period, followed by motorcycles. In terms of gasoline consumption, light-duty vehicles come third. Finally, diesel-powered HGVs occupy last place in terms of energy consumption. Fig. 7 shows annual fuel consumption for land transport in cubic meters.

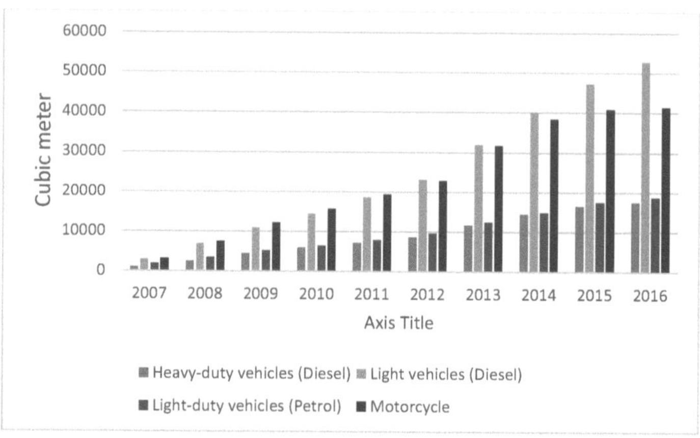

Figure 7 Annual fuel consumption for land transport in m³.
Source : TCNTCC (2020)

In Chad, less than 5% of the national road network is paved. These are mainly tracks vulnerable to climatic effects such as wind erosion, water erosion and silting. An effort in this area, towards more paved roads, would make the network more resilient to future climate change. (TCNTCC, 2020).

The increase in the number of vehicles and the resulting rise in fuel consumption have also triggered a rise in energy demand. Thus, fuel consumption for land transport is based on a number of simplifying assumptions, taking into account the realities of the country.

Road density, like population density, varies greatly according to climatic zone, ranging from a density of 6.4km/1000km^2 in the Saharan zone to 27.2km/1000km^2 in the Sahelian zone and 40.5km/1000km^2 in the Sudanian zone, as shown in Table 2:

Table 2 Road density by climatic zone

Zone Climate	Number of	Area	National and regional	Density (km/1000 km)^2

regions		km²	%	km	%	
Saharienne	3	600 350	46,8	3 858	15,4	6,4
Sahelian	12	490 570	38,2	13 368	53,4	27,2
Sudanese	7	193 080	15,0	7 813	31,2	40,5
Chad	22	1284000	100			19,5

Source : TCNTCC (2020)

Since 2000, the priority road network has grown by 35%. The asphalt road network has grown from around 350 km in 2000 to 1,500 km by the end of 2010 and 2,300 km in 2016. This progress was made possible by the arrival in 2003 of funds from oil revenues, as well as increased funding from development partners. Over the past two decades, work intensified between 2006 and 2014, with a maximum of 450 km of roads opened to traffic in 2010, according to Chad's Ministry of Infrastructure and Equipment. The last few years, from 2015 to 2017, were marked by the sluggish economic situation currently facing the country. Fig. 8 shows the cumulative length of paved roads from 2000 to 2016.

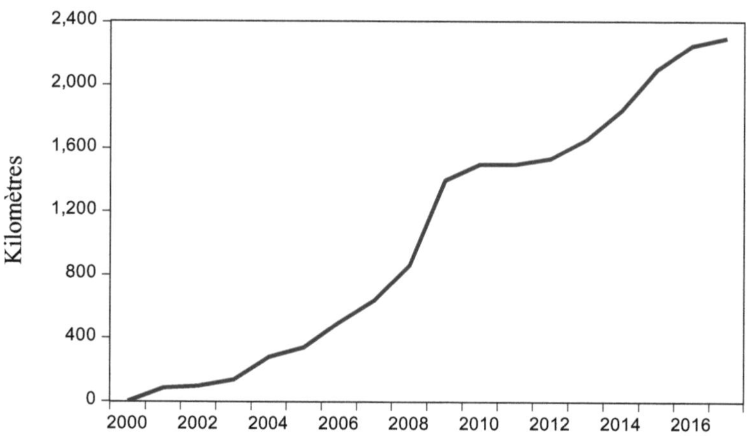

Figure 8 Cumulative length of paved roads from 2000 to 2016

The national road network comprises both paved and dirt roads. A large proportion of the dirt road network is in an advanced state of deterioration and passable only in the dry season. Dirt roads that are seasonally or semi-permanently passable (interrupted for less than a day in a row during the rainy season) used to benefit from maintenance or rehabilitation work. Intercity transport is almost exclusively by road (no railroads and very marginal river transport). Passenger transport is carried out by small vehicles (pick-ups or vans) and large vehicles (trucks converted into coaches). Maintaining a quality road network is difficult, given the size of the country, its low density (which means that the cost of the network is borne by few users), the mediocre quality of the geological materials available locally and the harshness of the climate.

During the rainy season, many towns are cut off from the rest of the world, and air travel becomes an alternative mode of transport. Mixed freight-passenger transport is very common, and a major factor in passenger comfort and safety. The most frequent causes of road fatalities are rollovers of mixed transport vehicles due to poor road quality, and crushing of passengers by overturned goods.

I.2. 2Distribution of petroleum products in land transport in Chad

The supply and distribution circuit for petroleum products includes refining, depot storage, transport and distribution. (ARSAT, 2022). The Djarmaya refinery supplies the entire country with petroleum products. But sometimes, with refinery maintenance generally causing shortages, particularly of butane gas, the Chad Downstream Petroleum Regulatory Authority, in agreement with the State, imports from Nigeria and Cameroon to meet national demand. (ARSAT, 2022).

These products are transported by road, pipeline and rail (ARSAT, 2022). Transporting petroleum products by road involves loading containers onto vehicles or filling tankers. It also involves transporting petroleum products from a point of shipment to a point of destination, and unloading containers from vehicles or emptying tank trucks. The construction and operation of pipelines must comply with the regulations in force and with general or local, fiscal, state, land, forestry, town planning and hygiene regulations, as well as with the technical standards and instructions issued by the Minister in charge of Petroleum. Rail transport of petroleum products involves loading containers onto wagons or filling tank wagons, moving petroleum products from a point of dispatch to a point of destination, and unloading containers from wagons or emptying tank wagons. Figure 9 below shows the distribution circuit.

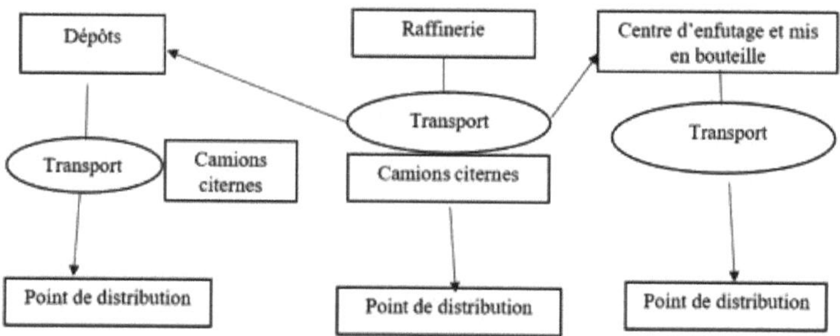

Figure 9 Marker distribution circuit diagram
Source : ARSAT (2022)

The Agence de Régulation du Secteur pétrolier Aval du Tchad does not manage oil depots, although they are the prerogative of the Société des Hydrocarbures du Tchad. However, some markers have their own depots. In addition, a number of national companies without the requisite skills in the oil industry are involved in transport, distribution and storage (storage in drums and/or tanks on concessions), but they are indispensable during periods of shortage.

In addition to the refinery, which plays a key role in supplying petroleum products to the national territory, it is accompanied by more than 70 markers (ARSAT, 2022). The most important marker here is undoubtedly Total Marketing Tchad, a French multinational that ranks among the super majors. Total Marketing is the leading distributor of petroleum products in Chad, and owns a large number of service stations: over twenty, in four provinces and one depot. On the other hand, Libya Oil Tchad seems to be overlooked, even though it is a major player, fully integrated in the downstream sector and effective in the oil industry in Chad with its takeover of Tamoil Tchad.

In Chad, there are over 255 service stations selling petroleum products in the country's 32 cities. The city of

N'Djamena is the most doped in terms of service stations, with over 155, or more than half of the country's total. It is followed by Abeché with 25 service stations, and Moundou comes third with 22. (ARSAT, 2022). Table 3 shows the geographical distribution of service stations in Chad.

Table 3 Geographical distribution of service stations in Chad

N°	Cities	Numbers
1	Abeché	25
2	Amdjarass	2
3	Ati	3
4	Bagassola	1
5	Biltine	1
6	Bisney	1
7	Bokoro	2
8	Bowl	4
9	Bongor	4
10	Doba	3
11	Faya	2
12	Goré	1
13	Guelendeng	4
14	Guereda	1
15	Iriba	1
16	Karmé	1
17	Kelo	2
18	Koundoul	2
19	Kournari	1
20	Mao	4
21	Massaguet	4
22	Massakory	2
23	Mongo	2
24	Moundou	22
25	Moyto	1

26	N'Djaména	147
27	N'Djaména Bilala	1
28	Ngoura	2
29	Ngouri	5
30	Oumhadjar	2
31	Sarh	1
32	Tiné	1
Total		255

Source : ARSAT (2022)

I.3 Justification of the parameters used

I.3. 1Justification of the parameters emissions of CO_2 emissions and consumption of petroleum products in the road sector

Countries' adherence to international agreements is a necessary condition for reducing global warming, but a sufficient condition is the implementation of climate change policies aimed at reducing fossil fuel consumption, as these are not neutral and influence energy consumption and emissions. CO_2 (Li et al., 2017; Solaymani et al., 2015).

Under normal atmospheric conditions, the gas mixture is composed of 78% nitrogen, 20% oxygen, 0.93% argon, 0.03% carbon dioxide and the remainder neon, helium, methane and hydrogen. The state of the air in the atmosphere can be said to be contaminated when the composition of the air is changed. Changes in the composition of incoming air can occur due to the inclusion of substances, energy and/or other components resulting from a potentially polluting activity (Sawitri et al., 2017). The main pollutants produced by motor vehicles that can cause air pollution include dust, carbon monoxide, carbon dioxide, hydrocarbons, nitrogen oxides and sulfur oxides (Qiao et al., 2015; Miller and Spoolman, 2011).

Carbon emissions from motor vehicles are one of the causes of global warming, since most of the contributing emissions are carbon dioxide (CO_2) plus methane (CH_4), nitrogen monoxide (N_2O), hydrogen per-fluoride, sulfur oxide, nitrogen oxide, per-fluorocarbons (PFCs), hydro-fluorocarbons (HFCs) and hydrogen sulfide. According to the United Nations Framework Convention on Climate Change, in addition to the six types of greenhouse gas, there are other types of greenhouse gas, including carbon monoxide (CO), nitrogen oxides (NOx), chlorofluorocarbons (CFCs) and non-volatile organic metal gases.

The three main types of gas known as greenhouse gases are CO_2 the CH_4O N_2O because they are considered as a layer of gas that acts as a heat-wave trap and the concentration in the atmosphere continues to increase until it is doubled (CHANGE, 2007). According to calculations made in recent years, the contribution of CO_2 in global warming has reached over 60%.

Emissions of CO_2 emissions are rising, reaching 27.0 billion tonnes of CO_2 in 2004, an average increase of 1.6% per year. Contributors to CO_2 emissions are distributed as follows: the United States (21.9% of total world emissions of CO_2 emissions), followed by China (17.4%) and India (4.1%). Indonesia accounts for 1.2% of global emissions. CO_2. The main source of CO_2 emissions comes from the use of fossil fuels (74% of total emissions worldwide), followed by the land convention (24%) and the cement industry (3%) (Sawitri et al., 2017).

Despite numerous international crises, Africa has experienced rapid growth averaging 5% per year over the past decade. This rapid growth has led to more conspiracies about Africa, and thus to its notorious transformation from a continent of insecurity, war and hunger to one of confidence in business and development opportunities. According to Cotte Poveda and Pardo Martínez (2011), Guo et al (2018) and Zhang et al. (2018)the use of energy actually promotes opportunities in the economy, minimizes the cost of travel and further stimulates the industrial sector, modernizing the economy. In 2015, Africa produced 8.1% of the world's energy, compared with 7.8% in 1971 (Mensah et al., 2019).

Despite this, energy consumption on the African continent remains low compared to other regions, according to the International Energy Agency in 2017. In 2009, for example, Europe's energy consumption was 11 times higher than that of sub-Saharan Africa, even though sub-Saharan African countries have larger populations (Mensah et al., 2019). It is therefore natural to expect that, with economic modernization, demographic changes and accelerating urbanization, Africa's energy consumption in the short and medium term will increase considerably.

In addition, the growing threat of global warming and climate change resulting from sustained economic development leading to greenhouse gas emissions poses a serious challenge to the world, and Africa is no exception. Carbon dioxide emissions from fossil fuels are considered a major source of global warming. As a result, the heavy dependence of most African countries on fossil fuels has become commonplace in this modern civilization. With rising greenhouse gas emissions, this modern practice has been documented as having a significant impact on the environmental and internal balance of the world's natural ecosystem.

Chad has been consuming large quantities of petroleum products for more than a decade, and air pollution through emissions of CO_2 emissions. It is therefore appropriate to examine the relationship between emissions of CO_2 emissions in Chad and petroleum product consumption in general.

Transport also emits CO_2 the most important greenhouse gas (GHG), and if global warming exceeds the safe threshold of 2°C, then the consequences could be bad and catastrophic (Pachauri and Meyer, 2014). In fact, there is already evidence that the safe threshold could actually be 1.5°C (Santos, 2017). To keep global warming below 2°C (or 1.5°C), atmospheric GHG concentrations need to be stabilized, which will eventually require zero net annual emissions (Santos, 2017).

The Paris Agreement has highlighted the need to ensure that the transport sector is not neglected in the effort to reduce global greenhouse gas (GHG) emissions. The Intergovernmental Panel on Climate Change (Pachauri and Meyer, 2014) stated that, while there are many paths to mitigating global warming, all of them require a substantial reduction in emissions of CO_2.

The transport sector as a whole is a major contributor to global warming, accounting for around 8 GtCO_2 or 24% of global CO_2 emissions (Teter, 2020). Of the total emissions, three quarters of the emissions reported by the transport sector came from the road transport sector. Worryingly, over the period 2000-2017, transport emissions rose steadily in all regions, despite the various global economic crises.

In Malaysia, the same scenario is playing out. The transport sector is the second largest GHG emitter, accounting for 20% of Malaysia's GHG emissions. Within this 20%, road transport is again the biggest culprit, responsible for a disproportionate 18% of GHG emissions EMISSIONS (MESTECC, 2018). Also, transport accounts for almost 30% of all carbon dioxide emissions in the EU

(European Union), 72% of which is road transport (Kubáňová et al., 2021). The European Union has therefore set itself the target of reducing transport-related emissions of CO_2 emissions from transport by 2050, compared with 1990 levels. This is an important part of the effort to reduce greenhouse gas emissions (Kubáňová et al., 2021)..

What's more, the transport sector is one of the main contributors to rising energy consumption and greenhouse gas emissions (Pita et al., 2020). Typically, the mode of transport includes road, air, water and rail (Pita et al., 2020). In many developed countries, such as Japan, Europe, Germany and North America, rail is the preferred mode of transport. (Pita et al., 2020; Hansen et al., 2016; Pita et al., 2020). In developing countries, on the other hand, road transport plays a major role in the overall transport sector (Pita et al., 2020).

Transport development is a key factor in improving living standards in urban areas and acts as a catalyst for economic development (Martínez-Jaramillo et al., 2017). However, transport is also a critical factor leading to resource depletion and a range of environmental problems (Ahanchian and Biona, 2014). In particular, road transport consumes most of the world's petroleum products (Duan et al., 2021) and is a major contributor to local air quality and global climate change (Duan et al., 2021; L. Zhang et al., 2019; Du et al., 2017).

Road traffic is the main contributor to the transport sector's carbon footprint, and reducing transport has become one of the main objectives of sustainable transport policy (Sobrino and Monzon, 2014).. Road transport is responsible for CO_2 emissions (Danish et al., 2018). Consequently, it is necessary to implement climate change policies throughout the economy, particularly in the road transport sector, which contributes the most to emissions of CO_2.

Before implementing a policy, it is necessary to identify the main drivers of carbon emissions in this sector. Consequently, to provide insight for policy-makers in implementing meaningful policies with specific targets for achieving the low-carbon objective in the transport sector. To do this, it is necessary to analyze the trend and characteristics of petroleum product consumption in the transport sector and carbon emission policies, and to break down the main drivers of carbon emissions in this sector. CO_2 emissions in this sector.

Diesel and gasoline are the main sources of CO_2 emissions in Chad. Given their higher consumption by the road transport sector, it becomes imperative to study the existing relationship between emissions of CO_2 emissions and consumption of petroleum products in the road transport sector. This involves identifying the factors related to road transport that directly or indirectly affect the growth in emissions of CO_2 emissions, and which accounts for the largest share of fuel consumption in Chad.

Fig. 10 shows the evolution of CO_2 emissions in Chad over the period 2008-2019. It is clear that CO_2 emissions in Chad over this period are increasing, and have accelerated significantly since 2011, when the NDJAMENA refinery came on stream. This growth can certainly be explained by the increased consumption of petroleum products in Chad and the fall in their prices. Emissions of CO_2 emissions in Chad peaked in 2015 and will not reach this level again until 2019. Chad therefore confirms the presence of emissions of CO_2 emissions, and a study of this kind is entirely appropriate in this context.

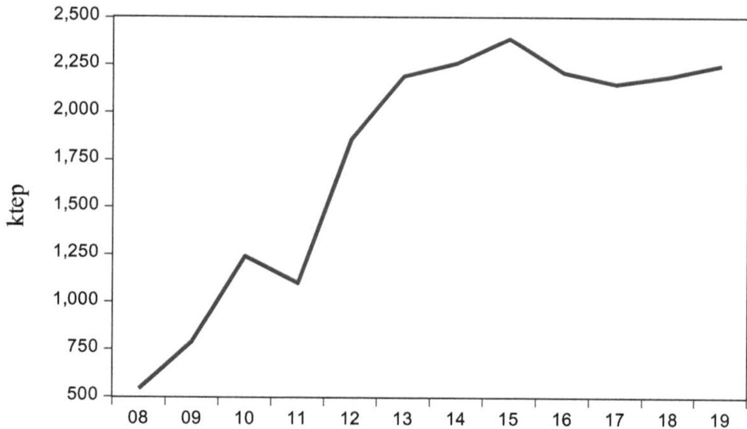

Figure 10 Evolution of CO_2 emissions in Chad from 2008 to 2019

Fig. 11 below, on the other hand, shows the consumption of petroleum products in Chad over the period 2008-2019. It increases slightly between 2008 and 2010. Consumption of petroleum products in Chad increased exaggeratedly in 2011, when the NDJAMENA refinery came on stream. The latter has led to a considerable drop in the price of petroleum products in Chad, which justifies this increased growth. The consumption of petroleum products in Chad has remained virtually constant over time, and its use as an exogenous variable is therefore justified.

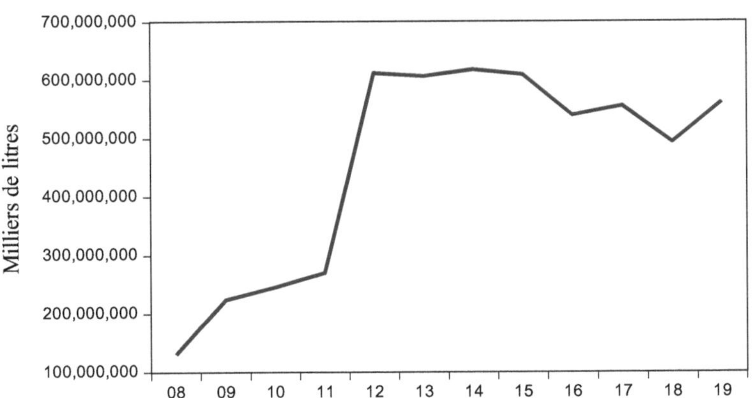

Figure 11 Consumption of petroleum products in Chad from 2008 to 2019

I.3.2Justification of oil price and urbanization parameters

Energy prices play the role of a rationing device in the management of energy services as part of the phenomenon of energy governance, and can therefore affect environmental quality (Shan et al., 2021; Fatima et al., 2021). In this respect, higher energy prices may lead to a governance struggle for energy efficiency technologies to improve environmental sustainability (Shan et al., 2021; Jabeen et al., 2021). However, lower energy prices are not conducive to such a struggle, and therefore worsen environmental quality (S. Zhang et al., 2019). Also, the introduction of fuel price reforms in a country can help raise domestic oil prices, which in turn can help reduce local oil consumption and pollutant emissions (Mahmood et al., 2020).

As energy has become an essential part of the economy in virtually every sector, oil has also become a strategic commodity for economies around the world. Arouri et al (2012) argue that while the nature of the link may change over time, crude oil plays an important role in all economies. As global oil prices fluctuate over time due to factors such as consumption of petroleum products, emissions of CO_2 emissions and many other factors, Africa remains one of the main regions in which revenues are most at risk. For example, most African oil-producing countries obtain 70-80% of national income from oil exports, while 35-60% of the GDP of the region's producers depends on oil (Mensah et al., 2019). These statistics show that dependence on oil for the economic survival of most African countries is high.

Understanding the link between energy consumption (fossil fuels), carbon emissions and the price of oil in this context in Chad, an oil-producing country, is therefore essential for policy-makers and economists.

Fig. 12 shows the evolution of gasoline and diesel prices in Chad over the period 2008-2019. The trend in petroleum product prices at the pump shows a steady rise in diesel and gasoline prices from 2008 to 2010. During this period, the petroleum products consumed in Chad came directly from neighboring countries, which explains the high prices of the various products. With the commissioning of the Chad Refining Company in 2011, pump prices for gasoline and diesel have fallen. In 2011, gasoline prices rose gradually, while diesel prices remained constant until 2014. The other years will be marked by increasing, decreasing and constant trends, which could be due to fluctuating oil prices on the international market and the policy of stabilizing petroleum product prices implemented by the Chadian government via ARSAT.

In addition, urbanization generally includes economic urbanization, population urbanization, land urbanization and social urbanization, and has an impact on emissions of CO_2 (Li and Haneklaus, 2022; S. Wang et al., 2018).

Urbanization is a major factor due to the existence and presence of humans and disrupts the global carbon cycle. Consequently, it can have a significant impact on natural resources and the environment, and can profoundly alter ecosystems (Grodzicki and Jankiewicz, 2022; Ulucak and Khan, 2020; Zhang et al., 2014).

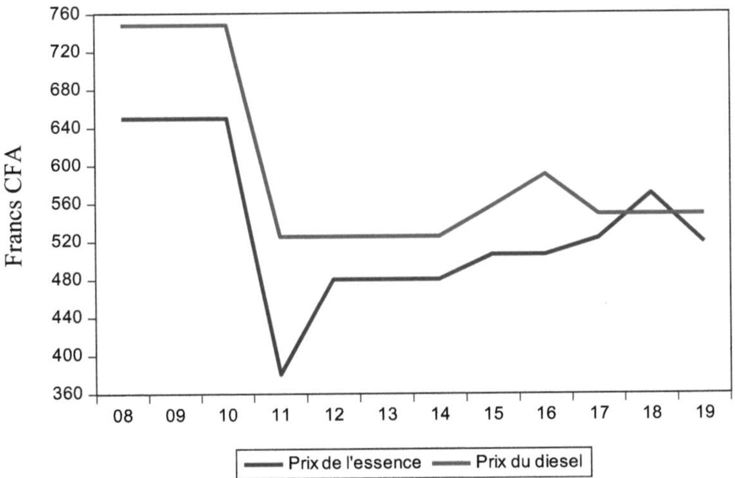

Figure 12 Gasoline and diesel price trends in Chad from 2008 to 2019

In general, an increase in the urban population leads to higher emissions of CO_2emissions, while an ageing population can reduce CO_2 emissions (Grodzicki and Jankiewicz, 2022).. Studying the dynamic link between urbanization and carbon emissions is useful for better understanding how targets for reducing CO_2 emissions can be achieved. Such a study is also useful for providing case studies from the most industrialized economies to industrializing countries (Li and Haneklaus, 2022).

Worldwide, urban areas are responsible for 71% of global energy-related emissions. CO_2 emissions. This percentage will rise to 76% by 2030, according to the International Energy Agency. High levels of urbanization, rapid economic growth and high energy emissions CO_2 emissions are characteristic of the African continent. Moreover, half of the population lives in urban areas on the African continent, and forecasts indicate an urbanization rate of 75% by 2050. This high urbanization rate has resulted in environmental degradation (Hossain, 2011). In Africa, the appropriate urbanization model is contrary to the requirements of a green, low-carbon economy. Thus, proper planning of urbanization has become necessary to limit carbon emissions. CO_2.

Chad is home to a population of nearly 15 million in 2018 (CDN, 2021). The average annual population growth rate rose from 2.4% for the period 1985-2000 to 3.9% for the period 2000-2015 (TCNTCC, 2020). This population is expected to reach 19.34 million in 2025 and 44.21 million in 2050 according to the trend scenario (TCNTCC, 2020). In view of the projected increase in Chad's population, it is becoming necessary to control it in order to take care of local pollution. Finally, several other authors in the literature have used the urbanization factor as an explanatory variable for emissions of CO_2. These include Mahmood et al (2022)and those of Shahbaz et al. (2014)those of Poumanyvong and Kaneko (2010) and those of Zhang and Cheng (2009).

Fig. 13 shows the evolution of urbanization in Chad between 2008 and 2019. It can be seen that the urban population in Chad over the period is growing. It is increasing quite considerably over time and deserves particular attention.

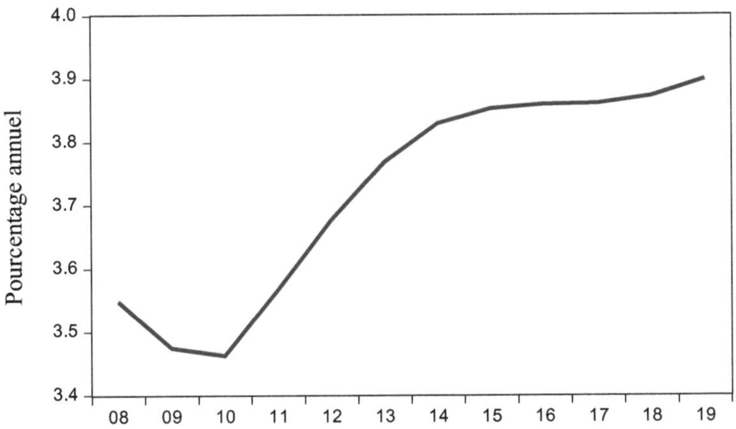

Figure 13 Urban population growth in Chad from 2008 to 2019

I.4   Literature review

I.4.1Relationship   between energy consumption in the transport sector and CO CO_2

A number of authors have provided a variety of studies on the relationship between energy consumption in the transport sector and greenhouse gas emissions. CO_2. There is therefore a wide range of analyses of this associated relationship in various countries. For example, Arora and Kaur (2020) examine the long-term relationship between fossil fuel consumption, economic growth and emissions of CO_2 emissions in the BRICS economies. The data covers the period 1990-2014. Causality results are obtained using the Dumitrescu-Hurlin panel causality technique. Thus, the authors found a unidirectional causality running from fuel consumption to environmental degradation. They conclude that governments in emerging economies need to focus more on non-conventional energy sources to keep the environment healthy while making growth sustainable in the long term.

Following the United Nations Sustainable Development Goals, Umar et al (2021) study the impact of biomass energy consumption, fossil fuel energy consumption and economic growth (GDP) on carbon dioxide emissions (CO_2) emissions in the U.S. transportation sector. This study also uses Gregory-Hansen cointegration, Hatemi-J cointegration, cointegration regression (FMOLS, DOLS and CCR) and the Breitung-Candelon Spectral causality test for the period between 1981Q1 and 2019Q4. The results reveal that the increase in fossil fuel energy consumption is associated with an increase in emissions of CO_2 emissions from the transport sector. In the long term, fossil fuel energy consumption leads to emissions of CO_2 emissions from the transportation sector in the USA at different frequency levels. This study offers policy perspectives for the transportation sector in the U.S. and in other similar economies that reproduce the same conditions.

Gokmenoglu and Sadeghieh (2019) examine the relationship between financial development and environmental degradation in Turkey from 1960 to 2011, using a multivariate framework that focuses on economic growth and fuel consumption as additional determinants of environmental degradation. The results show that, in the long term, fuel consumption has a positive and elastic impact on carbon emissions of 2.82%. Consequently, the error correction term implies that the CO_2 moves towards its long-term equilibrium level at an adjustment speed of 16.97% through the contributions of not only fossil fuel consumption, but also gross domestic product (GDP) and financial development. In symmetrical analyses, Wang et al (2021) showed a cointegration between oil dependency, emissions and income. The countries analyzed in this study, including China and India, have a high demand for oil and also have a high emission rate. CO_2 emissions. However, this cointegration is corroborated in countries such as France and the USA. The results suggest that oil-exporting countries with high emissions should CO_2 diversify their oil circuits and devise more comprehensive energy policies, capable of devising long-term action plans and helping to maintain a balance between oil demand and greenhouse gas emissions. CO_2. They also suggested supporting the market penetration of renewable energies to reduce these countries' oil dependency and make their economic and industrial activities cleaner.

Chandran and Tang (2013) evaluate the impact of transport sector energy consumption and foreign direct investment on emissions of CO_2 emissions for the 5 Asian economies using the cointegration method and Granger causality. The results show bidirectional causality between transport sector energy consumption, foreign direct investment and emissions of CO_2 emissions in Thailand and Malaysia. These authors propose the control of energy consumption in the transport sector as a solution to significantly reduce emissions of CO_2. They suggest that an efficient transport system and policies aimed at minimizing fossil fuel consumption should be put in place to promote environmental improvement.

The work of Solís and Sheinbaum (2013) presents a disaggregation of fuel consumption and related CO_2 emissions for passenger and freight road transport in Mexico. The results show that private gasoline-powered vehicles accounted for 32.6% of emissions of CO_2 emissions in 2010, followed by light freight gasoline vehicles with 25%, intercity diesel buses with 11.3% and heavy freight diesel vehicles with 12%. They suggest that mitigation of GHG emissions must be based on policies to reduce the fuel consumption of gasoline-powered vehicles, notably through fuel efficiency standards, but also on reducing the use of private cars by public transport and light freight logistics.

Schipper et al (2011) analyzed GHG emissions from the US transportation sector between 1960 and 2008 through economic growth, modal shift, efficiency and carbon content. In parallel, changes in transportation energy consumption due to fuel use and fuel-based economics are presented by Schipper et al. (2011) for several industrialized countries. Trends in transport emissions in Spain, in the context of the European Union, are analyzed by de Cáceres and Martínez (2009).

Studies on the relationship between energy consumption and CO CO_2 emissions, such as those by Antonakakis et al (2017) show that causality between CO_2 emissions and energy consumption is bidirectional, thus supporting the feedback hypothesis. Nnaji et al. (2013) study the causal relationship between electricity supply, fossil fuel consumption, emissions of CO_2 emissions and economic growth in Nigeria over the period 1971-2009 in a multivariate framework. Using the cointegration test approach, they found a short- and long-term relationship between the variables, with a positive and statistically significant relationship between emissions of CO_2 emissions and fossil fuel consumption. Among other things, they propose to reduce CO_2 emissions by developing alternatives to fossil fuel consumption, the main source of CO_2. Lotfalipour et al (2010) study the causal relationships between economic growth, emissions of CO_2 emissions and fossil fuel consumption using the Toda-Yamamoto method in Iran over the period 1967-2007. They use total consumption of fossil fuels, petroleum products and natural gas as three indicators of energy consumption. The empirical results show that there is no

Granger causality between total fossil fuel consumption and long-term carbon emissions. Consequently, reducing energy consumption appears to be an active means of reducing carbon emissions. CO_2. Alam et al. (2011) analyzed the relationship between energy consumption, GDP and CO_2 emissions in India using dynamic modeling. Ozcan (2013) examined the relationship between carbon emissions, energy use and GDP using panel data analysis. Zhang and Nian (2013) studied the influences of transport sectors on carbon emissions from a CO_2 emissions from a regional perspective in China, and quantified future emissions of CO_2.

Other authors have examined the relationship between fossil fuel consumption, CO CO_2 emissions, incorporating either the price of oil or population density as a proxy for urbanization. For example, Mensah et al. (2019) study the causal link between economic growth, fossil fuel consumption, carbon emissions and the price of oil over the period from 1990 to 2015, using a panel of 22 African countries. The results show a two-way causal link between fossil fuel consumption and carbon emissions. CO_2 emissions in the long and short term for all panels.

Uzair Ali et al (2022) studied the impact of economic development, fossil fuel consumption and population density on emissions of CO_2 emissions in India, Pakistan and Bangladesh using annual data over the period 1971-2014. The results show that fossil fuel consumption and population density have a positive impact on emissions of CO_2 emissions in the long term. On the other hand, the authors found that there is no causality between fossil fuel consumption and emissions of CO_2 . However, there is a short-term causality between population density and fossil fuel consumption, and emissions of CO_2 emissions influence population density. The authors suggest the use of efficient, low-carbon technologies.

Applying the autoregressive distributive lag model (ARDL) and the vector error correction model (VECM) over the period 1990-2015, Danish et al. (2018) examined the relationship between energy consumption in the transport sector, economic growth and carbon dioxide emissions (CO_2) emissions from the transport sector in Pakistan, taking into account foreign direct investment and urbanization. The empirical results of this research indicate a significant impact of transport sector energy consumption on carbon dioxide emissions. CO_2. Furthermore, the impact of urbanization on transport sector CO_2 emissions is statistically insignificant. They therefore suggest that the government should focus on promoting energy-efficient means of transport in order to improve environmental quality while reducing negative effects on economic growth.

Hossain (2011) examined the relationship between CO_2 emissions, energy consumption, economic growth, trade openness and urbanization for a panel of nine newly industrialized countries that included Malaysia, the Philippines and Thailand. The study shows that income and energy consumption have a significant long-term impact on greenhouse gas emissions in the Philippines and Thailand. CO_2 emissions in the Philippines and Thailand, but not in Malaysia. The Granger panel causality test indicates that there is no long-term causality between income, energy consumption and CO CO_2.

The oil sector and the price of oil could affect emissions of CO_2 emissions in oil-importing and oil-exporting countries. In their symmetric analysis, Mensah et al. (2019) explored the issue for the African continent using a panel cointegration approach. They used panel data from 1990 to 2015 for 22 African countries, and the sample contained both oil-exporting and non-exporting countries. For both categories, a one-way causal relationship was found between oil price and income, fossil fuel use and emissions of CO_2. This causality between various oil price indicators and pollution has also been found in other studies such as those by Ji et al. (2018) and Hammoudeh et al. (2014).

The work of Mahmood et al (2022) show that in the long term, higher oil prices have a positive impact on emissions of CO_2. It also has a negative effect, corroborating the technical and compositional effects in Kuwait. On the other hand, lower oil prices have a positive impact on emissions of CO_2 emissions in Bahrain and a negative effect in Kuwait.

Sadorsky (2009) states that rising oil prices have had a negative effect on renewable energy consumption in G7 countries. This finding implies that higher oil prices and oil market volatility can lead to an increase in CO_2 emissions by reducing renewable energy consumption. This result has a strong implication, particularly for oil-exporting countries, to control and regulate oil prices and ensure that the use of fossil fuels is reduced so that the negative effects on the environment are not irreversible. As part of a spatial analysis, Li et al. (2020) studied energy prices and their impact on emissions of CO_2 emissions in China. They showed that energy prices could have a significant negative impact on emissions of CO_2 emissions, both directly and indirectly, and that these energy prices have knock-on effects.

Böhringer et al (2021) argued that national emissions of CO_2 emissions could be reduced by a uniform economy-wide emissions pricing policy, and that revenues from CO_2 could be recycled in Germany. To achieve its environmental objectives, Germany has set higher prices for emissions of CO_2 emissions, which translates into higher household revenues. M. Y. Malik et al. (2020) analyzed oil prices, incomes and their impact on emissions of CO_2 emissions in Pakistan. For them, in the short term, oil prices have accelerated emissions of CO_2. However, in the long term, the relationship is reversed and oil prices reduce emissions of CO_2.

Ignoring the asymmetrical analysis in the GCC countries, the literature has probed the role of oil prices on emissions of CO_2. Aldubyan and Gasim (2021) analyzed the role of energy price reforms in Saudi Arabia and explored demand response and environmental and economic impacts. They reported that lower energy prices lead to increased energy demand and consumption. However, there is also a caveat, as there is an inevitable waste of relevant energy and resources with rapidly increasing demand. This dynamic leads to less focus on energy efficiency in general, let alone renewable or sustainable energies. Consequently, regulating energy prices can, in turn, have direct and indirect effects on greenhouse gas emissions. CO_2. If energy prices are kept at a competitive level, its use can be efficient, ensuring that environmental objectives are also maintained.

Another study on oil prices and emissions in the CO_2 emissions in the GCC region provides practical guidance on the subject. Alkathery and Chaudhuri (2021) studied oil prices, emissions and energy actions in the GCC region. They concluded the Co-mobility of oil prices, emissions and global clean energy production. They confirmed the spillover effects of these variables and market volatilities, and suggested revenue diversification policies.

Among other important determinants, urbanization is also a crucial determinant of emissions, and some works have investigated the link between urbanization and pollutant emissions within the framework of symmetrical analyses. Mahmood et al (2022) stipulate that urbanization positively affects emissions of CO_2 emissions in Bahrain, Oman, Qatar and the UAE (United Arab Emirates). They therefore suggest that these countries should impose a carbon tax on energy-intensive urban activities to discourage polluting emissions, and that the revenue from this tax should be used to encourage cleaner energy use in urban areas.

Meng et al (2021) argue that urbanization and industrialization both increase carbon emissions. They also claim that urbanization increases emissions faster than industrial growth. Their findings concluded that, while focusing on industrial development and rapid urban activity, countries need to estimate the pros and cons and potential environmental damage to the region, as the cost can sometimes be higher than the benefit, which can leave permanent after-effects. Indeed, the cost can sometimes be greater than the benefit, which can have a permanent negative impact on the environment. The results of the study confirm the findings of many other analyses carried out on a similar subject, which have shown that industrialization and urbanization appear to harm the environment (Dong et al., 2019; Al-Mulali et al., 2015; Cherniwchan, 2012).

S. Wang et al. (2018) examined the effect of urbanization, income level and other variables contributing to the increase in emissions of CO_2. They divided countries into several categories based on their income levels, which provided another level of context to their analysis. They suggest that CO_2 emissions are energy-related, at all income levels. Consequently, whether a country is low-income, middle-income or high-income, particular attention needs to be paid to the speed of urbanization in order to reduce emissions of CO_2.

Abdallh and Abugamos (2017) using a semi-parametric approach, explored a similar idea for the MENA (Middle East and North Africa) region and mentioned that emissions increase with urbanization and rising incomes. They suggested policies to reduce the negative impacts of growth and urbanization. In their view, this can be achieved in a number of ways, including stricter regulation of energy use and the adoption of more advanced, environmentally friendly technologies to achieve economic growth and urbanization while keeping the environment clean and sustainable.

Mahmood et al (2020) studied urbanization, the oil sector and CO CO_2 emissions and found a positive effect of urbanization and income on CO_2 emissions in Saudi Arabia. As a country with an abundance of oil and the largest oil exporter in the Middle East, Saudi Arabia's oil sector has significant effects on its economy and other socio-economic indicators. Consequently, there is no doubt that the country's oil sector will inevitably have a level of CO_2 emissions and may have significant negative effects on the environment. Furthermore, Majeed et al (2021) found that oil abundance has improved and urbanization has deteriorated the environment in the GCC (Gulf Cooperation Council) region.

Finally, the growth of China's road transport sector has led to a considerable increase in oil demand and emissions in the country over the past two decades, and these trends are likely to continue in the absence of specific measures to reduce the average carbon intensity of road vehicles. CO_2 emissions in the country over the past two decades, and these trends are set to continue in the absence of specific measures to reduce the average carbon intensity of road vehicles. With this in mind, the work of Gambhir et al. (2015 describes a model that enables a scenario analysis to be undertaken on the impact in terms of costs and emissions of CO_2 of replacing current vehicle transmission types with alternatives over the period 2010-2050. In addition, they undertake a detailed breakdown of the additional costs and emissions savings of CO_2 emissions savings of each low-carbon vehicle type into its component parts to calculate the marginal cost of reducing each vehicle and transmission type in 2050. The results of this study indicate that passenger cars and heavy-duty vehicles account for the majority of the potential for reducing emissions of CO_2 emissions in the future, but that, using central cost assumptions, alternative powertrains are significantly more cost-effective for trucks than for passenger cars. In the low-carbon scenario, demand for petroleum products (gasoline and diesel) is more than 40% lower than in the business-as-usual scenario in 2050. The total cost of mitigation in 2050 is $64 billion (US2010) per year, or 1.3% of total annual expenditure on road transport in China in 2050, using a 5% discount rate to annualize vehicle purchase costs, although this cost increases with higher discount rates. A sensitivity analysis shows that additional measures to those in the low-carbon scenario could further reduce emissions, in some cases at a negative cost. The availability and transparency of the model means that a range of other scenarios and sensitivities can be tested and developed, to help plan an optimal decarbonization strategy for this carbon-intensive sector.

Other authors such as Tsiakmakis et al. (2017) indicate that to assess the evolution of CO_2 emissions from road vehicles are generally based on estimates of the future composition of the fleet or on approaches that consider vehicles at a rather aggregated level. Thus, their work proposes a new method for the detailed calculation of emissions of CO_2 emissions from the European light-vehicle fleet, in order to obtain better results when projecting emissions of CO_2 emissions and to better support future policies. The results show that increases in emissions of CO_2 emissions are greater for cars with lower NEDC (New European Driving Cycle) emission values (29 and 25g CO_2/km for vehicles emitting 100 and 119g CO_2/km respectively). What's more, at higher emission levels, the NEDC and WLTP (Worldwide Light duty vehicle Test Procedure) show comparable results. We also have the work of Fontaras et al. (2017) which examines the influence of the various factors affecting fuel consumption and vehicle emissions, both on the road and in the laboratory. CO_2 emissions on the road and in the laboratory. Factors such as vehicle configuration and traffic conditions are again confirmed as having a major influence. Also, quality controls of the CO_2 emissions certification procedure, combined with in-use consumption monitoring, could be used to assess the gap on an ongoing basis. Finally, considering that the introduction of zero-emission vehicles and lightweight materials into a conventional, steel-intensive internal combustion engine fleet will have an impact on energy consumption and automotive material requirements, González Palencia et al (2012) have developed a dynamic bottom-up accounting model of the light vehicle fleet to reduce energy consumption and emissions of CO_2. The results show that the passenger car fleet in Colombia will increase sixfold between 2010 and 2050. Battery-powered electric vehicles offer the greatest reductions in energy consumption and greenhouse gas emissions. CO_2.

The above-mentioned studies focus in general on the impact of fossil fuel consumption, oil prices, urbanization and the vehicle fleet on greenhouse gas emissions. CO_2 emissions, either individually or grouped together; and the causal links between the different variables under study. Moreover, the results of these studies are inconclusive. Using Chad's time series, this study is the first to simultaneously relate emissions of CO_2 emissions, petroleum product consumption, petroleum product prices, the vehicle fleet and urbanization in general and in Chad in particular.

I.4.2 Forecasting models for emissions of CO_2 and a brief overview of related previous work

A variety of factors have uncertain and complex impacts on final carbon emissions, making it more difficult to forecast the emissions of CO_2 emissions from fuel combustion. To solve this problem, many published articles have attempted to forecast emissions of CO_2 emissions by sector. For example, Köne and Büke (2010) projected emissions of CO_2 emissions from fuel combustion in selected countries using trend analysis. Pérez-Suárez and López-Menéndez (2015) have combined Kuznets environmental projections with logistical growth models, to forecast emissions of CO_2 emissions for different countries. To forecast CO_2 emissions at provincial level, Sun et al. (2017) put forward a new extreme learning machine based on particle swarm optimization (PSO-ELM) by integrating factor analysis. Ding et al. (2019) reported that China's carbon emissions will peak in 2030 and then decline significantly under both low and high carbon tax scenarios. Wang et al. (2020) forecast that over the period 2019-2030, US carbon emissions will maintain a downward trend, while carbon emissions from China and India will continue to rise. Xu et al. (2019) have estimated that China's peak CO_2 emissions will occur in 2029, 2031 or 2035 at the level of 10.08, 10.78 and 11.63 billion tonnes in the low-growth, moderate-growth and high-growth scenarios. Based on Kaya ID, Yuan et al. (2014) provided a forecast and concluded that China's emissions will peak in 2030 and 2035 with 9.2 - 9.4 billion tonnes of CO_2. Yu et al (2018) estimated that China is likely to peak in 2023, with total emissions of 11.21 - 11.56 billion tonnes. Q. Wang et al. (2018) found that the annual growth rate of Chinese energy demand from 2017 to 2026 will be 1.36% to 1.70% (simple linear), 1.04% to 1.49% (hybrid linear), 1.80% to 2.34% (non-linear), respectively

From a conceptual point of view, these approaches can be classified into three categories (W. Xie et al., 2021). These include statistical models, machine learning methods and Grey models.

a) Statistical models

Since then, statistical models have been widely used in a variety of disciplines, due to the advantages of high explanatory power. To predict emissions of CO_2, Zhao and Du (2015) constructed an empirical panel regression model that can be seen as an extension of the Environmental Kuznets Curve (EKC) literature, and projected the total emissions of the sample countries in four scenarios as a function of the country's per capita emissions using multiple linear regression and multiple polynomial regression. Hosseini et al. (2019) explored carbon emissions by 2030 under business-as-usual and Sixth Development Plan assumptions. A. Malik et al. (2020) used the ARIMA (Autoregressive Integrated Moving Average) model to forecast Pakistan's CO_2 emissions from Pakistan in the China-Pakistan economic corridor. Belbute and Pereira (2020) used a fractionally integrated autoregressive moving average model to forecast emissions of CO_2 emissions from fossil fuel combustion and cement production in Portugal. Clearly, these models can achieve satisfactory results and present desirable explanatory power for a sufficient number of observations. However, their main limitation lies in the fact that they rely heavily on parameter estimation to identify relationships between variables.

Furthermore, statistical analysis models, including univariate models: logistic equation (Meng and Niu, 2011), trend analysis (Köne and Büke, 2010) and multivariate models: regression analysis (Piecyk and McKinnon, 2010) and (Meng et al., 2012)have also been widely used to predict energy-related CO_2 emissions.

b) Machine learning methods

As a type of advanced intelligent algorithm, machine learning methods can solve problems of CO_2 with the advantage of being able to perform complex calculations. Qiao et al (2020) have taken into account the stability neglected in previous literature and proposed a new hybrid algorithm by combining a genetic algorithm and lion swarm optimization to improve the performance of traditional least squares. The results of forecasting carbon dioxide emissions from developed countries have shown the merits of stronger global optimization capability and higher accuracy. Taking into account multiple influencing factors, Zhao et al. (2018) proposed a hybrid approach of the salp swarm algorithm and a least-squares support vector machine by PSO-optimizing model parameters. Results revealed that both improved predictive ability and reliability. Wen and Cao (2020) proposed a new method for forecasting energy-related carbon emissions in Shanghai in which they combined a support vector machine, an improved with integrated chaotic mutation and nonlinear weight index and principal component analysis. To forecast national carbon emissions in the building sector, Hong et al. (2018) investigated an optimized gene expression programming model based on meta-heuristic algorithms. Ameyaw et al. (2019) designed a long-short memory algorithm to predict emissions of CO_2 emissions from fossil fuel combustion alone, and the model proved effective in the absence of exogenous variables and assumption requirements. The main drawback of machine learning methods is the need for a considerable amount of data. On the other hand, if the sample size is small, performance will be poor. What's more, the internal operating mechanism is unknown.

In addition, intelligent non-linear models, including artificial neural networks (Kaynar et al., 2011)fuzzy regression (Azadeh et al., 2009) and a support vector machine (Sun and Liu, 2016) have been established to predict CO_2 with high efficiency. Artificial neural networks are analytical techniques modelled on learning processes inspired by biological systems, in particular that of the human cognitive system and the neurological functions of the brain. Neural networks are distributed information processing systems made up of numerous simple computing elements interacting via weighted connections. Inspired by the architecture of the human brain, neural networks have certain characteristics, such as the ability to learn from examples and to capture subtle functional relationships between data, even if the underlying relationships are unknown and difficult to describe. An artificial neural network is generally made up of a large number of computing elements called nodes. Neural networks are simply parameterized non-linear functions that can be fitted to data for predictive purposes. They are therefore data-driven, self-adaptive methods. There has been considerable interest in developing artificial neural networks to solve a wide range of problems in different fields. A major application of artificial neural networks is forecasting (Ardakani et al., 2018; Zhang et al., 2016). Another flexible computational technique that is a very powerful method for building a complex, non-linear relationship between input and output data is the fuzzy rule-based approach. A specific approach in neuro-fuzzy development is ANFIS (Adaptive Network-based Fuzzy Inference System), which combines the learning capabilities of the neural network and the reasoning capabilities of fuzzy logic to give improved prediction capabilities compared to using a single methodology (Jang, 1993). Mardani et al. (2018) have applied an adaptive neuro-fuzzy inference system model to study the relationship between CO_2 emissions and energy consumption in G20 countries. The results show that as energy consumption increases in some countries, emissions of CO_2 emissions will continue to rise.

c) Grey models

As introduced by Ofosu-Adarkwa et al (2020)it is best to focus on a time sequence that involves the most relevant shock, when attempting to model a time series impacted by shocks. In short, we should reduce the sequence closest to the current time to capture shocks. In this context, adequate information on annual carbon emissions is not always available. The grey prediction model is known as an effective approach for forecasting the development trend of an uncertain system with sparse data. Grey prediction models offer an alternative modelling technology, providing accurate projections even in the presence of sparse samples. Grey prediction theory, founded by Deng, J.L. (1982)aims to determine the optimal system parameters of a grey differential equation, by which dynamic system behavior can be accurately reproduced and predicted. As a result, Grey's forecasting models have been widely used in a variety of problems, such as foreign currency exchange rates (Kayacan et al., 2010)power load (Wu et al., 2015a)environmental pollution (Pai et al., 2015)unemployment rates (Chen and Huang, 2013)policy effectiveness (Chen et al., 2015) and energy consumption (Hamzacebi and Es, 2014a; Xiong et al., 2014).

Forecasting emissions of CO_2 emissions from fuel combustion is essential if policy-makers are to identify ready-to-use targets for effective reduction plans, and to improve energy policies and plans. This is how W. Xie et al (2021) propose the continuous conformable nonlinear fractional Grey Bernoulli model as a new method for forecasting the future evolution of emissions of CO_2 emissions from fuel combustion in China. This model predicts China's future emissions of CO_2 emissions from fuel combustion in China by 2023, representing 10039.80 million tonnes. Ofosu-Adarkwa et al (2020) applied Grey's forecasting method to estimate emissions of CO_2 emissions from the Chinese cement industry. The study provided a reference to help policy-makers meet emission reduction targets.

Guo et al (2020) implemented a pre-treatment Grey model to analyze energy savings in China. The results highlight that green development not only encourages economic and industrial developments, but also significantly reduces energy consumption. Pao et al (2012) used Grey's predictive model to forecast China's CO_2 emissions, energy consumption and economic growth in China. The results show that China needs to improve its energy efficiency to reduce energy waste. Xie et al (2021) have proposed a new, robust, reweighted multivariate Grey model to predict emissions of CO_2 emissions in

European Union member countries over the period 2010 to 2016. Hu et al (2021) proposed an optimized Grey prediction model to predict carbon dioxide emissions for twenty countries.

Dengiz et al (2018) predicted the carbon dioxide emissions of seven developed countries by applying Grey's forecasting method. The results indicated that seven selected countries need to take action to reduce emissions of CO_2 . Wu et al. (2015) implemented a new multi-variable Grey model to forecast urban population, energy consumption and emissions of CO_2 emissions in Brazil, Russia, India, China and South Africa. The results show that these countries need to improve their energy efficiency to reduce emissions of CO_2. Hamzacebi and Karakurt (2015) applied a Grey prediction model to forecast energy-related CO_2 emissions in Turkey. The results showed that Turkey's CO_2 emissions will increase in 2025 compared to 2010.

Ding et al (2017) have proposed a new multivariate Grey model to predict emissions of CO_2 emissions from fuel consumption in China. The results showed that the growth in emissions of CO_2 emissions in China is due to high energy consumption. Wang and Ye (2017) verified the non-linear and uncertain link between emissions of CO_2 emissions and GDP, and improved a non-linear Grey model to predict China's future CO_2 emissions from China from 2014 to 2020 under three assumptions: high, medium and low economic growth in China.

It is clear from the above studies that several models have been applied to predict emissions of CO_2 emissions from petroleum product consumption in the road transport sector, such as statistical models, machine learning models and Grey's models, which stand out from the others. Grey models are very simple to understand and have several advantages (Sapnken and Tamba, 2022) they only require a minimum of four observations, there's no need to make heavy assumptions about the nature of these observations, and they are capable of making very precise forecasts (Sapnken, 2023). In addition to this, variables such as petroleum product prices, vehicle fleet and urbanization were used as additional input data. This study therefore uses the Sequential GMC (1, N) - GA model derived from the traditional Grey model, which, to our knowledge, is being implemented more and more for forecasting emissions of CO_2 emissions, due to their efficiency and simplicity, based on the consumption of petroleum products in the road transport sector in general.

Conclusion

This research therefore focuses on the linear and causal relationship between CO_2 emissions and the consumption of petroleum products (gasoline and diesel), in addition to the other determinants of CO_2 emissions, i.e. the vehicle fleet, the price of oil and population growth as a proxy for urbanization. It also looks at future emissions of CO_2 emissions, based on petroleum product consumption, petroleum product prices, vehicle fleet and urbanization. The gaps in the literature that this study aims to fill are: (i) the non-existence of studies modeling the emissions of CO_2 emissions due to the consumption of petroleum products in road transport in Chad. (ii) the scarcity of studies forecasting future emissions of CO_2 emissions in Chad. This work has therefore enriched the literature concerning, on the one hand, the linear and causal relationship that would exist between emissions of CO_2 emissions, petroleum product consumption, vehicle fleet, oil price and urbanization, which has not been covered to date, and on the other hand, the forecasting of

CO_2 emissions in Chad. The results of this study will guide Chadian oil pollution management decision-makers in the adoption of policies related to the effects of petroleum product consumption in Chad's transport sector.

CHAPTER II
METHODOLOGICAL ASPECTS AND DATA SOURCES

Introduction

This chapter describes the approach used to determine the causal links between emissions of CO_2 emissions, the consumption of petroleum products (Super and Gasoil), the vehicle fleet, the price of petroleum products at the pump (Super and Gasoil) and urbanization. Most studies agree that urban population growth is a proxy for urbanization. For the vehicle fleet, they generally use the combination of cars, trailers and related equipment. This chapter also looks at the process used to forecast emissions of greenhouse gases. CO_2. Most forecasting models allow us to establish the functional link (if any) between a so-called dependent variable and several other so-called independent variables. In our study, the dependent variable is emissions from CO_2 emissions, the independent variables are respectively the consumption of petroleum products (Super and Gasoil), the vehicle fleet, the price of petroleum products at the pump (Super and Gasoil) and urbanization.

II.1 Establishing relationships between series

II.1. 1Linear regression models

The simple regression model can be used to study the relationship between two variables. For reasons we shall see, the simple regression model has limitations as a general tool for empirical analysis. Nevertheless, it is sometimes appropriate as an empirical tool. Interpreting the simple regression model is good practice for studying multiple regression, which we'll do next.

a) Simple regression model

i. Types of regression models

- Linear regression

$$Y_i = \beta_0 + \beta_1 X_i + \varepsilon_i$$
(1)

- Y_i represents the result of the dependent variable (response) for the i[ème] experimental/sampling unit
- X_i represents the level of the independent variable (predictor) for the i[ème] experimental/sampling unit
- $\beta_0 + \beta_1 X_i$ represents the linear (systematic) relationship between Y_i and X_i (zero conditional mean)
- β_0 represents the mean of Y when X=0 (Y intersection)
- β_1 represents the change in the mean of Y when X increases by 1 (slope)
- ε_i represents the random error term

Note that β_0 and β_1 are unknown parameters. We estimate them using the least-squares method.

a) Polynomial regression (non-linear)

$$Y_i = \beta_0 + \beta_1 X_i + \beta_2 X_i^2 + \varepsilon_i$$
(2)

This model allows for a curvilinear relationship (as opposed to a straight line). Both linear and polynomial regressions are prone to problems when Y predictions are made outside the range of X values used to fit the model. This is known as extrapolation.

ii. Least squares estimation

The parameter estimation algorithm is as follows:

1. Obtain a sample of n pairs (X_1 , Y_1) .... (X_n , Y_n).
2. Plot the Y values on the vertical axis (up/down) against the corresponding X values on the horizontal axis (left/right).
3. Select line $\hat{Y}_i = b_0 + b_1 X_i$ which minimizes the sum of the squares of the vertical distances from the observed values (Y_i) to their adjusted values ($\hat{Y}_i$).

Note: $SCR = \sum_{i=1}^{n}(Y_i - \widehat{Y}_i)^2$

4. b$_0$ is the Y-intercept for the estimated regression equation.

5. b$_1$ is the slope of the estimated regression equation

iii. M easurements of variation

- S omme des carrés

- Total sum of squares = Regression sum of squares + Error sum of squares
- Total variation = Explained variation + Unexplained variation
- Total sum of squares (Total variation):

$$SCT = \sum_{i=1}^{n}(Y_i - \bar{Y})^2 \quad df_T = n - 1 \tag{3}$$

- Sum of squares of the regression (Variation explained):

$$SCE = \sum_{i=1}^{n}(\widehat{Y}_i - \bar{Y})^2 \quad df_R = 1 \tag{4}$$

- Sum of squared errors (unexplained variation):

$$SCR = \sum_{i=1}^{n}(Y_i - \widehat{Y}_i)^2 \quad df_E = n - 2 \tag{5}$$

- Coefficients of determination and correlation

- Determination coefficient

- Proportion of variation in Y "explained" by regression on X

$$r^2 = \frac{Variation\ expliquée}{Variation\ totale} = \frac{SCE}{SCT} \quad 0 \leq r^2 \leq 1 \tag{6}$$

- Correlation coefficient

- Measuring the direction and strength of the linear association between Y and X

$$r = sign(b_1)\sqrt{r^2} \qquad -1 \le r \le 1$$

(7)

- Standard error of estimate (Residual standard deviation)

- Data standard deviation estimation ( )$V(Y_i) = V(\varepsilon_i) = \sigma^2$

$$S_{YX} = \sqrt{\frac{SSE}{n-2}} = \sqrt{\frac{\sum_{i=1}^{n}\left(Y_i - \hat{Y}_i\right)^2}{n-2}}$$

(8)

iv. Model assumptions

- Normally distributed errors
- Heteroskedasticity (constant error variance for Y at all levels of X)
- Independent errors (generally verified when data are collected over time or space)

v. Residual analysis

Residues: $e_i = Y_i - \hat{Y}_i = Y_i - (b_0 + b_1 X_i)$ (9)

vi. Graphics

- Plotting e_i as a function of $\hat{Y}_i$ can be used to verify a linear relationship, a constant variance.
  - If the relationship is non-linear, a U-shaped pattern appears
  - If the error variance is not constant, a funnel-shaped model appears
  - If the hypotheses are met, a point cloud appears
- Plot e_i as a function of X_i can be used to verify a linear relationship, a constant variance
  - If the relationship is non-linear, a U-shaped model appears
  - If the error variance is non-constant, a funnel-shaped model appears

- If the hypotheses are met, a cloud of random points appears.
- Plotting e_i as a function of i can be used to check independence when data are collected over time.
- Histogram of e_i
  - If the distribution is normal, the histogram of residuals will be mound-shaped, around 0

vii. Measuring autocorrelation - Durbin-Watson test

a) Plot residuals against time
- If the errors are independent, there will be no model (random cloud centered at 0).
- If the errors are not independent (dependent), we expect them to be close to each other over time (distinct curve shape, centered on 0).
b) Durbin-Watson test
- H_0 : Errors are independent (no autocorrelation between residuals)
- H_A : Errors are dependent (positive autocorrelation between residuals)
- Test statistics or t-stat:

$$D = \frac{\sum_{i=2}^{n} (e_i - e_{i-1})^2}{\sum_{i=1}^{n} e_i^2} \tag{10}$$

- Decision rules (The values of $d_L(k,n)$ and $d_U(k,n)$ are given in Wooldridge (2010):
- If $D \geq d_u(k,n)$ then conclude H_0 (independent errors) where k is the number of independent variables (k=1 for simple regression)
- If $D \leq d_L(k,n)$ then conclude H_A (dependent errors) where k is the number of independent variables (k=1 for simple regression)
- If $d_L(k,n) \leq D \leq d_U(k,n)$ in this case refrain from any judgment (we may need a longer series).

Note: This is a test that you "want" to include in favor of the null hypothesis. If you reject the null hypothesis in this test, a more complex model must be fitted.

ix. Inferences concerning slope

a. t-test

The test is used to determine whether the population-based slope parameter (β_1) is equal to a predetermined value (often, but not necessarily 0). Tests can be one-sided (predetermined direction) or two-sided (either direction).

b. Two-tailed t-test

$$H_0:\beta_1 = \beta_1^0 \qquad H_A:\beta_1 \neq \beta_1^0$$

$$TS:t_{obs} = \frac{b_1 - \beta_1^0}{S_{b_1}} \qquad S_{b_1} = \frac{S_{YX}}{\sqrt{\sum_{i=1}^{n}(X_i - \overline{X})^2}}$$

$$RR:|t_{obs}| \geq t_{\frac{\alpha}{2},n-2}$$

$$(11)$$

c. One-tailed t-test (upper tail, signs reversed for lower tail)

$$H_0:\beta_1 = \beta_1^0 \qquad H_A:\beta_1 > \beta_1^0$$

$$TS:t_{obs} = \frac{b_1 - \beta_1^0}{S_{b_1}} \qquad S_{b_1} = \frac{S_{YX}}{\sqrt{\sum_{i=1}^{n}(X_i - \overline{X})^2}}$$

$$RR:t_{obs} \geq t_{\frac{\alpha}{2},n-2}$$

$$(12)$$

d. F-test (based on k independent variables)

A test based directly on the sum of squares that tests the specific hypotheses of whether the slope parameter is 0 (two-tailed). The book describes the general case of k predictor variables, for simple linear regression, k=1

$$H_0: \beta_1 = 0 \qquad H_A: \beta_1 \neq 0$$

$$TS: F_{obs} = \frac{MCR}{MCE} = \frac{SCR/k}{SCE/(n-k-1)}$$

78

$$= F_{\alpha,k,n-k-1} \qquad \text{RR: } F_{obs} \qquad (13)$$

e. Analysis of variance (based on k predictor variables)

Table 4 Analysis of variance (based on k predictor variables)

Source	df	Sum of squares	Medium square	F
Regression	k	SCE	MCE=SCE/k	F =MCR/MCE$_{obs}$
Error	n-k-1	SCR	MCR=SCR/(n-k-1)	---
Total	n-1	SCT	---	---

f. $(1-\alpha)100\%$ Confidence interval for the slope parameter, β_1

$$b_1 \pm t_{\frac{\alpha}{2},n-2} S_{b_1}$$

$$(14)$$

- If the integer interval is positive, conclude $\beta_1 > 0$ (Positive association)
- If the interval contains 0, conclude (do not reject) $\beta_1 = 0$ (No association)
- If the whole interval is negative, conclude $\beta_1 < 0$ (Negative association)

g. Estimation of the mean and prediction of individual results

** Estimation of average population response when X=X$_i$

Parameter: $E[Y|X = X_i] = \beta_0 + \beta_1 X_i$ (15)

Estimated points: $\hat{Y}_i = b_0 + b_1 X_i$ (16)

** $(1-\alpha)100\%$ Confidence interval for $E[Y|X = X_i] = \beta_0 + \beta_1 X_i$:

$$\hat{Y}_i \pm t_{\frac{\alpha}{2},n-2} S_{YX} \sqrt{h_i} \qquad h_i = \frac{1}{n} + \frac{(X_i - \overline{X})^2}{\sum_{i=1}^{n}(X_i - \overline{X})^2}$$

(17)

** Predict a future outcome when X=X$_i$

Future result: $Y_i = \beta_0 + \beta_1 X_i + \varepsilon_i$ (18)

Points prediction: $\hat{Y}_i = b_0 + b_1 X_i$ (19)

** $(1-\alpha)100\%$ Prediction interval for $Y_i = \beta_0 + \beta_1 X_i + \varepsilon_i$

$$\hat{Y}_i \pm t_{\frac{\alpha}{2},n-2} S_{YX} \sqrt{1+h_i} \qquad h_i = \frac{1}{n} + \frac{(X_i - \overline{X})^2}{\sum_{i=1}^{n}(X_i - \overline{X})^2}$$

(20)

x. Problems associated with regression analysis

- Knowledge of model assumptions
- Knowledge of methods for verifying model assumptions
- Alternative methods when model assumptions are not met

- Not understanding the theoretical context of the problem

The most important way of identifying problems with model assumptions is based on plots of data and residuals. Often, transformations of the dependent and/or independent variables can solve the problems. Other, more complex methods can also be used.

xi. Calculation formulas for simple linear regression

a) Normal equations (based on SCR minimization by

$$\sum_{i=1}^{n} Y_i = nb_0 + b_1 \sum_{i=1}^{n} X_i$$

$$\sum_{i=1}^{n} X_i Y_i = b_0 \sum_{i=1}^{n} X_i + b_1 \sum_{i=1}^{n} X_i^2$$

calculation)

$$\tag{21}$$

b) Formulas for calculating slope, b_1

$$b_1 = \frac{SCXY}{SCXX}$$

$$SCXY = \sum_{i=1}^{n}(X_i - \overline{X})(Y_i - \overline{Y})^2 = \sum_{i=1}^{n} X_i Y_i - \frac{(\sum_{i=1}^{n} X_i)(\sum_{i=1}^{n} Y_i)}{n}$$

$$SCXX = \sum_{i=1}^{n}(X_i - \overline{X})^2 = \sum_{i=1}^{n} X_i^2 - \frac{(\sum_{i=1}^{n} X_i)^2}{n}$$

$$\tag{22}$$

c) Calculation formula for y-intercept, b_0

$$b_0 = \overline{Y} - b_1 \overline{X}$$

$$\tag{23}$$

d) Formula for calculating the total sum of squares of the regression, SCT

$$SCT = \sum_{i=1}^{n}(Y_i - \overline{Y})^2 = \sum_{i=1}^{n} Y_i^2 - \frac{(\sum_{i=1}^{n} Y_i)^2}{n}$$

$$\tag{24}$$

e) Formula for calculating the sum of squares explained by the regression, SCE $SCE =$

$$\sum_{i=1}^{n}(\widehat{Y}_i - \overline{Y})^2 = b_0 \sum_{i=1}^{n} Y_i + b_1 \sum_{i=1}^{n} X_i Y_i - \frac{(\sum_{i=1}^{n} Y_i)^2}{n}$$

$$\tag{25}$$

f) Formula for calculating the sum of squares of the regression residuals, SCR

$$SCR = \sum_{i=1}^{n}(Y_i - \widehat{Y}_i)^2 = \sum_{i=1}^{n} Y_i^2 - b_0 \sum_{i=1}^{n} Y_i - b_1 \sum_{i=1}^{n} X_i Y_i \qquad (26)$$

g) Formula for calculating the standard error of the slope, S_{b_1}

$$S_{b_1} = \frac{S_{YX}}{\sqrt{\sum_{i=1}^{n}(X_i - \bar{X})}} \qquad (27)$$

b) Multiple linear regression

i. Multiple regression model with k independent variables

General case: We have k variables which we control or know in advance and which are used to predict Y, the response (dependent variable). The k independent variables are labelled $X_1\ X_2\ X_3 \dots X_K$. The levels of these variables for the ith case are denoted by $X_{1i}, \dots, X_{ki}$. Note that simple linear regression is a special case where k=1, so the methods used are just basic extensions of what we've done before.

$$y_i = \beta_0 + \beta_1 x_{1i} + \beta_2 x_{2i} + \cdots + \beta_k x_{ki} + \varepsilon_i \qquad (28)$$

where β_j is the variation in the mean of Y when the variable X_j increases by one unit, while holding the remaining k-1 independent variables constant (partial regression coefficient). We also speak of the slope of Y with the variable X_j holding the other predictors constant.

ii. Equation (prediction) fitted by least squares (obtained by SCR minimization):

$$\hat{Y}_i = b_0 + b_1 X_{1i} + \cdots + b_k X_{ki}$$

(29)

iii. Coefficient of multiple determination

Proportion of the variation in Y "explained" by the regression on the k independent variables.

$$R^2 = r_{Y.1,\ldots,k}^2 = \frac{\sum_{i=1}^{n}(\hat{Y}_i - \bar{Y})^2}{\sum_{i=1}^{n}(Y_i - \bar{Y})^2} = \frac{SCE}{SCT} = 1 - \frac{SCR}{SCT}$$

(30)

iv. Adjusted R²

Used to compare models with different sets of independent variables in terms of predictive ability. Penalizes models with unnecessary or redundant predictors.

$$R^2 = r_{Y.1,\ldots,k}^2 = \frac{\sum_{i=1}^{n}(\hat{Y}_i - \bar{Y})^2}{\sum_{i=1}^{n}(Y_i - \bar{Y})^2} = \frac{SCE}{SCT} = 1 - \frac{SCR}{SCT}$$

(31)

v. Residual analysis for multiple regression

Very similar to simple regression. The only difference is that residuals are plotted against each independent variable. Similar interpretations of graphs. A review of graphs and interpretations:

$$\text{Residuals: } e_i = Y_i - \hat{Y}_i = Y_i - (b_0 + b_1 X_{1i} + \cdots + b_k X_{ki}) \quad (32)$$

** Graphics:

- Plotting e_i as a function of $\hat{Y}_i$ can be used to verify a linear relationship, a constant variance
  - If the relationship is non-linear, a U-shaped pattern appears
  - If the error variance is non-constant, a funnel-shaped model appears

- If the assumptions are met, a cloud of random points appears.

- Plotting e_i as a function of X_{ji} for each j can be used to verify the linear relationship with respect to X_j
  - If the relationship is non-linear, a U-shaped pattern appears
  - If the assumptions are met, a random cloud of points appears
- Plotting e_i as a function of i can be used to check independence when data is collected over time.
  - If the errors are dependent, a smooth model appears

  - If the errors are independent, a cloud of random points appears.
- Histogram of e_i
  - If the distribution is normal, the histogram of residuals will be mound-shaped, around 0

vi. F-test for global model

Used to test whether one of the independent variables is linearly associated with **Y**

vii. **Analys e of** variance

Sum of total squares and df:

$$SCT = \sum_{i=1}^{n}(Y_i - \bar{Y})^2 \quad df_T = n - 1 \tag{33}$$

Sum of squares explained by the regression:

$$SCE = \sum_{i=1}^{n}(\widehat{Y}_i - \bar{Y})^2 \quad df_R = k \tag{34}$$

Sum of squared residuals:

$$SCR = \sum_{i=1}^{n}(Y_i - \widehat{Y_i})^2 \quad df_E = n - k - 1$$

$$(35)$$

Table 5 Analysis of variance

Source	df	Sum of squares	Medium square	F
Regression	k	SCE	MCE=SCE/k	F =MCR/MCE$_{obs}$
Error	n-k-1	SCR	MCR=SCR/(n-k-1)	---
Total	n-1	SCT	---	---

vii. F-test for global model

$H_0: \beta_j = 0$ (Y is not linearly associated with any of the independent variables)

$H_A: not\ all\ \beta_j = 0$ (At least one of the independent variables is associated with Y)

$$TS: \quad F_{obs} = \frac{MSR}{MSE} \qquad (36)$$

$$RR: \quad F_{obs} \geq F_{\alpha,k,n-k-1} \quad (37)$$

P value: Zone in the F distribution to the right of the F_{obs}

viii.   Inferences concerning individual regression coefficients

Used to test or estimate the slope of Y with respect to X_j, after controlling for all other predictor variables.

** t-test for β_j

$$H_0: \beta_j = \beta_j^{0} \quad \text{(Often, but not necessarily 0)} \quad (38)$$

$$H_A: \beta_j = \beta_j^{0} \tag{39}$$

$$\text{TS:} \quad t_{obs} = \frac{b_j - \beta_j^0}{S_{b_j}} \tag{40}$$

$$\text{RR:} \quad |t_{obs}| \geq t_{\frac{\alpha}{2}, n-k-1} \tag{41}$$

** P-value: Twice the area of the t distribution to the right of $| t |_{obs}$

ix. Confidence interval $(1 - \alpha)100\%$ for β_j

$$b_j \pm t_{\frac{\alpha}{2}, n-k-1} S_{b_j} \tag{42}$$

- If the integer interval is positive, conclude $\beta_j > 0$ (Positive association)
- If the interval contains 0, conclude (do not reject) $\beta_j = 0$ (No association)
- If the whole interval is negative, conclude $\beta_j < 0$ (Negative association)

x. Testing parts of the model

Consider two blocks of independent variables:

- The A block containing k-q independent variables
- Block B containing q independent variables

We wish to test whether the set of independent variables in block B can be removed from a model containing the set of independent variables. That is, the variables in block B do not add any predictive value to the variables in block A.

Notation:

- SCE(A&B) is the sum of squares of the regression for the model containing the two blocks
- SCE(A) is the sum of squares of the regression for the model containing block A
- MCR(A&B) is the mean square error for the model containing the two blocks
- k is the number of predictors in the model containing the two blocks
- q is the number of predictors in block B

H$_0$: The β in block B are all 0, since the variables in block A have been included in the model.

H$_A$: The β in block B are not all 0, since the variables in block A have been included in the model.

$$\text{TS: } F_{obs} = \frac{\left[\frac{SCE(A\alpha B)-SCE(A)}{q}\right]}{MSE(A\alpha B)} = \frac{\left[\frac{SCE(B|\,A)}{q}\right]}{MSE(A\alpha B)} =$$

$$\frac{MCE(B|\,A)}{MSE(A\alpha B)} \tag{43}$$

$$\text{RR: } F_{obs} \geq F_{\alpha,q,n-k-1} \tag{44}$$

** P-value: Zone in the distribution Y to the right of the F_{obs}

xi. Coefficient of partial determination (notation differs from text)

Measures the fraction of variation in Y explained by X_j , which is not explained by other independent variables.

$$r^2_{Y2.1} = \frac{SCE(X_1,X_2)-SCE(X_1)}{SCT-SCE(X_1)} = \frac{SCE(X_2|\,X_1)}{SCT-SCE(X_1)} \tag{45}$$

$$r^2_{Y3.12} = \frac{SCE(X_1,X_2,X_3) - SCE(X_1,X_2)}{SCT - SCE(X_1,X_2)} = \frac{SCE(X_3| X_1,X_2)}{SCT - SCE(X_1,X_2)}$$

$$(46)$$

$$r^2_{Yk.12\ldots k-1} = \frac{SCE(X_1,\ldots,X_k) - SCE(X_1,\ldots X_{k-1})}{SCT - SCE(X_1,\ldots,X_{k-1})} =$$

$$\frac{SCE(X_k| X_1,\ldots,X_{k-1})}{SCT - SCE(X_1,\ldots,X_{k-1})}$$

$$(47)$$

Note that since labels are arbitrary, they can be constructed for any independent variable.

xii. Quadratic regression models

When the relationship between Y and X is non-linear, it can often be approximated by a quadratic model.

Population model:

$$Y_i = \beta_0 + \beta_1 X_{1i} + \beta_2 X_{1i}^2 + \varepsilon_i$$

$$(48)$$

Adapted model:

$$\hat{Y}_i = b_0 + b_1 X_{1i} + b_2 X_{1i}^2$$

$$(49)$$

Test for all associations: $H_0: \beta_1 = \beta_2 = 0$ $\quad$ $H_0: \beta_1\ and/ or\ \beta_2 \neq 0$ $\quad$ (Test F)

Test for a quadratic effect: $H_0: \beta_2 = 0$ $\quad$ $H_0: \beta_2 \neq 0$ $\quad$ (t-test)

xiii. Models with dummy variables

Dummy variables are used to include categorical variables in the model. If the variable has m levels, we include $m - 1$ dummy variables. The simplest case is that of 2-level binary variables, such as gender. The dummy variable takes the value 1 if the characteristic of interest is present, 0 if it is absent.

xiv. Mod èle sans interaction

Consider a model with an independent numerical variable (X_1) and a dummy variable (X_2). Then the model for the two groups (feature present and feature absent) has the following relationships with Y:

$$Y_i = \beta_0 + \beta_1 X_{1i} + \beta_2 X_{2i} + \varepsilon_i \qquad E(\varepsilon_i) = 0$$

(50)

Characteristic present ($X_{2i} = 1$):

$$E(Y_i) = \beta_0 + \beta_1 X_{1i} + \beta_2(1) = (\beta_0 + \beta_2) + \beta_1 X_{1i}$$

Characteristic absent ($X_{2i} = 0$): $E(Y_i) = \beta_0 + \beta_1 X_{1i} + \beta_2(0) = \beta_0 + \beta_1 X_{1i}$

Note: the two groups have different ordinates, but the slopes with respect to X_1 are equal.

xv. Interaction model

This model allows the slope of Y in relation to X_1 to be different for the two groups in relation to the categorical variable:

$$Y_i = \beta_0 + \beta_1 X_{1i} + \beta_2 X_{2i} + \beta_3 X_{1i} X_{2i} + \varepsilon_i \qquad E(\varepsilon_i) = 0$$

(51)

$X_{2i} = 1$: $E(Y_i) = \beta_0 + \beta_1 X_{1i} + \beta_2(1) + \beta_3 X_{1i}(1) = (\beta_0 + \beta_2) + (\beta_1 + \beta_3) X_{1i}$

$X_{2i} = 0$: $E(Y_i) = \beta_0 + \beta_1 X_{1i} + \beta_2(0) + \beta_3 X_{1i}(0) = \beta_0 + \beta_1 X_{1i}$

Note: the two groups have different ordinates and slopes. More complex models can be fitted with multiple numerical predictors and dummy variables. See examples.

II.1.2 Transformations to meet the assumptions of the regression model

There is a wide range of transformation possibilities to improve the model when error assumptions are violated. We'll look specifically at two types of logarithmic transformations that work well in a wide range of problems.

Other transformations may be useful in very specific situations.

a) Logarithmic transformations

These transformations can often solve the problems of non-linearity, unequal variances and non-normality of errors. We'll look at logarithms in base 10, and natural logarithms.

$$X' = log_{10}(X) \qquad\qquad X = 10^{X'}$$

(52)

$$X' = ln(X) \qquad\qquad X = e^{X'}$$

(53)

Case 1: Multiplicative model (with non-negative multiplicative errors)

$$Y_i = \beta_0 X_{1i}^{\beta_1} X_{2i}^{\beta_2} \varepsilon_i$$

(54)

Taking the logarithms (in base 10) of the two sides, we obtain the model :

$$\log_{10}(Y_i) = \log_{10}(\beta_0 X_{1i}^{\beta_1} X_{2i}^{\beta_2} \varepsilon_i) = \log_{10}(\beta_0) + \log_{10}(X_{1i}^{\beta_1}) + \log_{10}(X_{2i}^{\beta_2}) + \log_{10}(\varepsilon_i) =$$
$$\log_{10}(\beta_0) + \beta_1 \log_{10}(X_{1i}) + \beta_2 \log_{10}(X_{2i}) + \log_{10}(\varepsilon_i)$$

Procedure: First, take the base-10 logarithms of Y, X_1 et X_2 and fit the linear regression model to the transformed variables. The relationship will be linear for the transformed variables.

Box 2: Exponential model (with non-negative multiplicative errors)

$$Y_i = e^{\beta_0 + \beta_1 X_{1i} + \beta_2 X_{2i}} \varepsilon_i$$

(55)

Taking the natural logarithms of the Y values, we obtain the model :

$$\ln(Y_i) = \beta_0 + \beta_1 X_{1i} + \beta_2 X_{2i} + \ln(\varepsilon_i)$$

(56)

i. Multicollinearity

Problem: in observational studies and models with polynomial terms, the independent variables may be highly correlated with each other. Two types of problem then arise. **Intuitively,** the variables explain the same part of the random variation in Y. This makes the partial regression coefficients seem insignificant for the individual terms (since they explain the same variation). **Mathematically,** the standard errors of the estimates are inflated, resulting in wider confidence intervals for the individual parameters, and smaller t-statistics for testing whether the individual partial regression coefficients are zero.

ii. Variance inflation factor (VIF)

$$VIF_j = \frac{1}{1-r_j^2}$$

$$(57)$$

Where r_j^2 is the coefficient of determination when the variable X_j is regressed on the remaining j-1 independent variables. There is a *VIF* for each independent variable. A variable is considered problematic if its VIF is equal to or greater than 10.0. In the case of multicollinearity, theory and common sense should be used to select the best variables to retain in the model. Complex methods such as principal component regression and ridge regression have been developed, but will not be pursued here, as they are of no use in explaining which independent variables are associated with and cause changes in Y.

b) Time series and forecasts

i. Classic multiplicative model

Time series can generally be broken down into components representing the overall level, linear trend, cyclical patterns, seasonal changes and random irregularities. Series that are measured at the annual level cannot highlight seasonal patterns. In the multiplicative model, the effects of each component enter multiplicatively.

ii. Multiplicative pattern observed at annual level

$$Y_i = T_i C_i I_i \tag{58}$$

where the components are trend, cycle and irregularity effects

iii. Multiplicative pattern observed at seasonal level

$$Y_i = T_i C_i S_i I_i \tag{59}$$

where the components are trend, cycle, season and irregularity effects

c) Smoothing a time series

Time series data is generally smoothed to reduce the impact of random irregularities in individual data points .

i. Mobile yennes

Centered moving averages are averages of the value of the current point and neighboring values (above and below the point). Because they are centered, they are generally odd. The 3- and 5-period moving averages at time point i would be as follows:

$$MA(3)_i = \frac{Y_{i-1} + Y_i + Y_{i+1}}{3} \qquad MA(5)_i = \frac{Y_{i-2} + Y_{i-1} + Y_i + Y_{i+1} + Y_{i+2}}{5} \tag{60}$$

Clearly, there can be no moving averages at either end of the series.

ii. Exponential smoothing

This method uses all the previous data in a series to smooth it, with exponentially decreasing weights on observations further back in time:

$$E_1 = Y_1 \qquad E_i = WY_i + (1-W)E_{i-1} \quad i = 2,3,\ldots \quad 0 < W < 1$$

(61)

The exponentially smoothed value for a current time point can be used as a forecast for the next time point.

$$\hat{Y}_{i+1} = E_i = WY_i + (1-W)\hat{Y}_i \qquad (62)$$

iii. Least-squares adjustment of trends and forecasts

We consider three trend models: linear, quadratic and exponential. In each model, the independent variable X_i is the observation period i. If the first observation is period 1, then $X_i = i$.

iv. Linear trend model

Population model:

$$Y_i = \beta_0 + \beta_1 X_i + \varepsilon_i \qquad (63)$$

Adjusted forecasting equation:

$$\hat{Y}_i = b_0 + b_1 X_i \qquad (64)$$

v. Quadratic trend model

Population model:

$$Y_i = \beta_0 + \beta_1 X_i + \beta_2 X_i^2 + \varepsilon_i \qquad (65)$$

Adjusted forecasting equation:

$$\hat{Y}_i = b_0 + b_1 X_i + b_2 X_i^2 \qquad (66)$$

vi. Exponential trend model

Population model:

$$Y_i = \beta_0 \beta_1^{X_i} \varepsilon_i \qquad \log_{10}(Y_i) = \log_{10}\beta_0 + X_i \log_{10}(\beta_1) + \log_{10}(\varepsilon_i)$$

$$(67)$$

Adjusted forecasting equation:

$$\hat{\log}_{10}(Y_i) = b_0 + b_1 X_i \qquad \hat{Y}_i = (10^{b_0})(10^{b_1})^{X_i}$$

$$(68)$$

The forecasting equation for this exponential model is obtained by first fitting a simple linear regression of the logarithm of Y on X, then putting the least squares estimates of the y-intercept and slope into the second equation.

vii. Model selection based on differences

First differences:

$$Y_i - Y_{i-1} \qquad i = 2,\ldots,n \tag{69}$$

Second differences:

$$(Y_i - Y_{i-1}) - (Y_{i-1} - Y_{i-2}) = Y_i - 2Y_{i-1} + Y_{i-2} \qquad i = 3,\ldots,n \tag{70}$$

Percentage differences:

$$\frac{Y_i - Y_{i-1}}{Y_{i-1}} \times 100\% \qquad i = 2,\ldots,n \tag{71}$$

If the first differences are approximately constant, then the linear trend model is appropriate. If the second differences are approximately constant, then the quadratic trend model is appropriate. If the percentage differences are approximately constant, the exponential trend model is appropriate.

d) Autoregressive model for trend adjustment and forecasting

Autoregressive models allow us to consider the average response as a linear function of the previous response(s).

i. 1st order model

Population model:

$$Y_i = A_0 + A_1 Y_{i-1} + \varepsilon_i \tag{72}$$

Adjusted equation:

$$\hat{Y}_i = a_0 + a_1 Y_{i-1} \tag{73}$$

ii. P model^{ième} order

Population model:

$$Y_i = A_0 + A_1 Y_{i-1} + \cdots + A_P Y_{i-P} + \varepsilon_i \tag{74}$$

Adjusted equation:

$$\hat{Y}_i = a_0 + a_1 Y_{i-1} + \cdots + a_P Y_{i-P} \tag{75}$$

iii. Significance test for the highest order term

$$H_0 : A_P = 0 \qquad H_A : A_P \neq 0$$

$$TS: t_{obs} = \frac{a_P}{S_{A_P}} \qquad RR: |t_{obs}| \geq t_{\frac{\alpha}{2}, n-2P-1} \tag{76}$$

This test can be used repeatedly to determine the order of the autoregressive model. You can start with a large value of P (say 4 or 5, depending on the length of the series), then continue fitting simpler models until the test for the last term is significant.

iv. Prediction equation for $P^{ième}$ adjusted order

Start by fitting the P-order autoregressive model to a data set containing n observations. Then we can obtain forecasts in the future as follows:

$$\hat{Y}_{n+1} = a_0 + a_1 Y_n + a_2 Y_{n-1} \cdots + a_p Y_{n-P+1}$$

$$\hat{Y}_{n+2} = a_0 + a_1 \hat{Y}_{n+1} + a_2 Y_n \cdots + a_p Y_{n-P+2}$$

$$\hat{Y}_{n+j} = a_0 + a_1 \hat{Y}_{n+j-1} + a_2 \hat{Y}_{n+j-2} \cdots + a_p \hat{Y}_{n+j-P}$$

$$(77)$$

In the third equation (the general case), $\hat{Y}$ are observed values when $j \le P$, and are forecast values when $j > P$.

e) Cho isir a forecasting model

To determine which of a set of potential forecast models to use, we first need to obtain the forecast errors (residuals). Two widely used measures are the mean absolute deviation (MAD) and the root mean square error (MSE).

$$e_i = Y_i - \hat{Y}_i \qquad MAD = \frac{\sum_{i=1}^{n} |Y_i - \hat{Y}_i|}{n} = \frac{\sum_{i=1}^{n} |e_i|}{n} \qquad MSE = \frac{\sum_{i=1}^{n} (Y_i - \hat{Y}_i)^2}{n} = \frac{\sum_{i=1}^{n} e_i^2}{n}$$

$$(78)$$

The plots of residuals versus time should be a random cloud, centered on 0. Linear, cyclical and seasonal patterns may also appear, leading to the fitting of models taking them into account.

f) Forecasting time series of quarterly or monthly series

Consider the exponential trend model, with quarterly and monthly dummy variables, based on series data.

i. Exponential model with quarterly data

We need to create three dummy variables: Q_1, Q_2, and Q_3 where $Q_j = 1$ if the observation took place in quarter j, 0 otherwise. Let X_i be the coded quarterly value (if the first quarter of the data is to be labeled as 1, then $X_i = i$). The population model and its transformed linear model are:

$$Y_i = \beta_0 \beta_1^{X_i} \beta_2^{Q_1} \beta_3^{Q_2} \beta_4^{Q_3} \varepsilon_i$$

$$\log_{10}(Y_i) = \log_{10}(\beta_0) + X_i \log_{10}(\beta_1) + Q_1 \log_{10}(\beta_2) + Q_2 \log_{10}(\beta_3) + Q_3 \log_{10}(\beta_4) + \log_{10}(\varepsilon_i)$$

$$(79)$$

Fit the multiple regression model with the logarithm to base 10 of Y_i as the response and X_i (which is usually i),

$$\log_{10}(\hat{Y}_i) = b_0 + b_1 X_i + b_2 Q_1 + b_3 Q_2 + b_4 Q_3$$

$$\hat{Y}_i = 10^{b_0 + b_1 X_i + b_2 Q_1 + b_3 Q_2 + b_4 Q_3}$$

Q_1 , Q_2 , and Q_3 as the independent variables. The fitted model is

$$(80)$$

ii. Exponential model with monthly data

A similar approach is adopted when the data is monthly. We create 11 dummy variables: M_1 ....M_{11} which take the value 1 if the month of observation is the month corresponding to the dummy variable. In other words, if the observation concerns the month of January, $M_1 = 1$ and all other M_j are equal to 0. The population model and its transformed linear model are as follows:

$$Y_i = \beta_0 \beta_1^{X_i} \beta_2^{M_1} \beta_3^{M_2} \cdots \beta_{12}^{M_{11}} \varepsilon_i$$

$$\log_{10}(Y_i) = \log_{10}(\beta_0) + X_i \log_{10}(\beta_1) + M_1 \log_{10}(\beta_2) + M_2 \log_{10}(\beta_3) + \cdots + M_{11} \log_{10}(\beta_{12}) + \log_{10}(\varepsilon_i)$$

$$(81)$$

$$\log_{10}(\hat{Y_i}) = b_0 + b_1 X_i + b_2 M_1 + b_3 M_2 + \cdots + b_{12} M_{11}$$

$$\hat{Y_i} = 10^{b_0 + b_1 X_i + b_2 M_1 + b_3 M_2 + \cdots + b_{12} M_{11}}$$

(82)

Equation (82) above is the fitted model.

II.2 Data sources

In our study, in order to determine the causal links that exist between emissions of CO_2 emissions, petroleum product consumption, vehicle fleet, petroleum product prices and urbanization, the annual data used in this research covers the period from 2008 to 2019 for Chad (i.e. 12 observations). The choice of this period was preferred in view of the availability of data in Chad, where we used CO_2 emissions as the endogenous or explained variable, and petroleum product consumption as the explanatory variable, which includes diesel and gasoline consumption. The consumption of diesel and gasoline was favored given their flexible form and their importance as petroleum products in Chad, as they appear to be the most widely consumed in road transport.

Thus, to better illustrate this relationship, in addition to the two variables of emissions of CO_2 emissions and petroleum product consumption, we have integrated other variables, namely the vehicle fleet, urbanization and the price of petroleum products. We have used urban population growth as a proxy for urbanization, and the vehicle fleet used here takes into account all passenger cars, vans, buses, rigids, tractors, semi-trailers, trailers and motorcycles. The data concerning CO_2 emissions come from the World Bank database and are expressed in kilotonnes.

Data on petroleum product consumption and prices come from Société de Raffinage au Tchad in N'Djamena, supplemented by data from ARSAT (2022) and are expressed in thousands of liters and local currency respectively. For the vehicle fleet, data come from the Ministry of Transport and Road Safety in Chad, and data concerning urban population growth in Chad as a proxy for urbanization come from the World Bank database (World Bank, 2022) and are expressed as an annual percentage.

II. 3Empirical model

The work of Danish et al (2018) uses energy consumption in the transport sector and urbanization to model the impact of the latter on greenhouse gas emissions. CO_2. In addition, the recent study by Mahmood et al. (2022) uses oil prices and urbanization in its model to study the impact of the latter on emissions of CO_2 emissions in the GCC (Gulf Cooperation Council) countries. Based on the above studies and drawing on the work of Solís and Sheinbaum (2013)those of Alkathery and Chaudhuri (2021) and Majeed et al (2021)This research focuses on the relationship between emissions of CO_2emissions, the consumption of petroleum products (Super and Gasoil), the vehicle fleet, the price of petroleum (Super and Gasoil) and urbanization in Chad, and proposes the following theoretical models:

$$CO_{2t} = f(CSU_t, PAU_t, PSU_t, URB_t) \tag{83}$$

$$CO_{2t} = f(CGA_t, PAU_t, PGA_t, URB_t) \tag{84}$$

Where CO_{2t}, CSU_t, CGA_t, PAU_t, PSU_t, PGA_t and URB_t respectively represent emissions of CO_2 emissions, supermarket consumption, diesel consumption, vehicle fleet, supermarket price, diesel price and urbanization respectively.

All variables were natural log-transformed following Ahmed et al. (2016) in equation (1) to obtain consistent and reliable empirical results (Shahbaz et al., 2016). The natural logarithm thus makes it possible to smooth the various variables and make interpretations in the form of elasticity. Empirical models between CO_2 and all other variables are as follows:

$$ln\,CO_{2t} = \alpha_0 + \alpha_1 lnCSU_t + \alpha_2 lnPAU_t + \alpha_3 lnPSU_t + lnURB_t + \mathcal{E}_t \qquad (85)$$

$$lnCO_{2t} = \alpha_0 + \alpha_1 lnCGA_t + \alpha_2 lnPAU_t + \alpha_3 lnPGA_t + \alpha_4 lnURB_t + \mathcal{E}'_t \qquad (86)$$

Where $ln\,CO_{2\,t}$, $ln\,CSU_t$, $ln\,CGA_t$, $ln\,PAU_t$, $ln\,PSU_t$, $ln\,PGA_t$ and $ln\,URB_t$ represent respectively the natural logarithm of the emissions of. CO_2 emissions, supermarket consumption, diesel consumption, vehicle fleet, supermarket price, diesel price and urbanization, respectively. α_0 the constant and $\mathcal{E}_t$, $\mathcal{E}'_t$ the error terms of equations (85) and (86) respectively.

II.3.1 STATIONARITY TEST Stationarity test and optimum criterion

Before processing a time series, its stochastic characteristics need to be studied. If these characteristics (expectation and variance) change over time, the time series is considered non-stationary; in the case of an invariant stochastic process, this implies that the series has no trend, no seasonality and, more generally, no factors that change over time. The time series is then said to be stationary.

In formal terms, the stochastic process y_t is stationary if :

- $E(Y_t) = E(Y_{t+m}) = \mu$ $\forall t$ and $\forall m$ the mean is constant and independent of time;
- $Var(Y_t) < \infty$ $\forall t$ the variance is finite and independent of time;
- $Cov(Y_t, Y_{t+k}) = E[(Yt - \mu)(Y_{t+k} - \mu)] = \gamma_t$, the covariance is independent of time.

Stationarity is a necessary condition to avoid spurious regressions, otherwise Ordinary Least Squares (OLS) estimators do not converge to the true coefficients, and the usual Student's (t) and Fisher's (f) tests are no longer valid, where the results could be significant when they are not. When using temporal data, it is essential that they retain a constant distribution over time. However, economic chronicles are rarely realizations of stationary random processes. Whether a process is stationary or not determines the choice of model to be adopted.

As a general rule, if the series under study stems from a stationary process, we look for the best model among the class of stationary processes to represent it, and then estimate this model. If, on the other hand, the series is the result of a non-stationary process, we must seek to "stationarize" it, i.e. find a stationary transformation of this process. We then model and estimate the parameters associated with the stationary component. The difficulty lies in the fact that there are different sources of non-stationarity, and that each source of non-stationarity has its own method of stationarization.

In the case of non-stationarity (of the deterministic/TS or random/DS type), the authors (Dickey and Fuller, 1979) suggested stationarization methods: first difference for non-stationary DS (Differencing Stationary) series, or deviation from trend for non-stationary TS (Trend Stationary) series.

a) TS process

It exhibits non-stationarity of a deterministic nature, and the most commonly used TS process is written as :

$$X_t = f(t) + \varepsilon_t \tag{87}$$

Where f(t) is a polynomial function of time, linear or nonlinear, and ε_t a stationary process. The simplest TS process is represented by a polynomial function of degree 1. The TS process is then called linear in this case and is written :

$$X_t = a_0 + a_1 t + \varepsilon_t \tag{88}$$

Where ε_t represents the model error at date t. It is white noise by definition stationary. The TS process is non-stationary because $E(x_t) = a_0 + a_1 t$ depends on time (t). By the ordinary least squares method, knowing the values a_0 and a_1, the process x_t can be stationaryized by subtracting the value of x_t in t from the estimated value $â_0 + â_1 t$. This example can be generalized to polynomial functions of any degree

b) DS process

As previously mentioned, there is another form of non-stationarity, arising not from the presence of a deterministic trend component, but from a stochastic source. Unlike TS processes, which are non-stationary in expectation, DS processes are non-stationary in variance "$V(x_t) = \sigma^2\varepsilon\ t$". The DS process is a process that can be made stationary by differentiation. This process is said to be of first order if:

$$X_t = \beta + X_{t-1} + \varepsilon_t \tag{89}$$

Where ε_t is white noise.

By introducing the constant β into the DS process, two different processes can be defined if :

- $\beta = 0$: the DS process is drift-free, written as follows:

$$X_t = X_{t-1} + \varepsilon_t \tag{90}$$

Since ε_t is white noise, the DS process is called a random walk. To stationarize this type of process, we rely on differentiation.

$$X_t = X_{t-1} + \varepsilon_t \rightarrow X_t - X_{t-1} = \varepsilon_t \rightarrow \Delta X_t = \varepsilon_t \tag{91}$$

- $\beta \neq 0$: the process is called DS with drift, and is written as follows:

$$X_t = \beta + X_{t-1} + \varepsilon_t \rightarrow X_t - X_{t-1} = \beta + \varepsilon_t \rightarrow$$
$$\Delta X_t = \beta + \varepsilon_t \tag{92}$$

In summary, to stationaryize a TS process, the correct method is ordinary least squares (time regression). For a DS process, the difference filter should be used. The choice of a DS or TS process as chronicle structure is therefore not neutral.

A non-stationary series is said to be integrated of order "d" if, after being differentiated d times, it is rendered stationary, and the series is said to have a unit root. There are two classes of unit root tests:

- Tests that adopt as H_o non-stationarity: we distinguish two cases in relation to the dynamics of the trend;

- The deterministic component follows a linear trend " the Dickey-Fuller test (DF), the Augmented Dickey-Fuller test (ADF) 1981 and the Phillips and Perron test (PP) 1988 "
- The deterministic component follows a non-linear trend "the test of Ouliaris, Park and Phillips 1989 and the test of Perron 1989".
- The tests that adopt H_0 Kwiatkowski, Phillips, Schmidt and Shin 1992 test".

Non-stationarity can be detected by visually observing time series graphs and correlograms, but Unit Root Tests not only detect the existence of non-stationarity, but also determine which non-stationarity it is (the TS or DS process).

In order to apply unit root tests, it is first necessary to choose the number of lags "p". The "p" lag value is determined using either the partial autocorrelation function, the Box-Pierce statistic, or the Akaike (AIC), Schwarz (SC) or Hannan-Quinn (HQ) criteria. The latter are used to determine the optimal offset. An optimal offset is one whose estimated model offers the minimum value of one of the listed criteria. Their values are calculated as follows:

$$\text{AIC(p)} = log|\Sigma| + \frac{2}{T} n^2 p$$

(93)

$$\text{SIC(p)} = log|\Sigma| + \frac{\log T}{T} n^2 p$$

(94)

$$\text{HQ(p)} = log|\Sigma| + \frac{2\log T}{T} n^2 p$$

(95)

Where: Σ = variance-covariance matrix of the estimated residuals; T = number of observations; p = lag of the estimated model and n = number of regressors.

The first step is to check the stationarity of the time series. For our purposes, the Augmented Dickey-Fuller (ADF) test and the Phillips and Perron test will be used, as they are easy to implement on the Eviews software we'll be using. Indeed, the results of these two tests are almost identical.

The most widely used test for testing the presence of the unit root and verifying stationarity on level and difference series is the Augmented Dickey-Fuller (ADF) 1981 test. The simple Dickey-Fuller test is based on the assumption that the errors are generated by a white noise process, yet there is no reason why, a priori, the error should be uncorrelated. The Augmented Dickey-Fuller (ADF) test takes into account the autocorrelation of errors (parametric correction of autocorrelation). The test consists of testing the null hypothesis H_0 : $\rho=0$ against the alternative H_1 : $|\rho| < 0$ and is based on ordinary least squares estimation of the following three models with ε_t is the error term :

- Trend and constant model

$$\Delta X_t = \rho\ X_{t-1} + c + bt + \sum_{m=1}^{p} \varphi_m\ \Delta X_{t-m} + \quad \varepsilon_t\ (96)$$

- Model without trend and with constant

$$\Delta X_t = \rho\ X_{t-1} + c + \sum_{m=1}^{p} \varphi_m\ \Delta X_{t-m} + \quad \varepsilon_t\ (97)$$

- Model with no trend and no constant

$$\Delta X_t = \rho\ X_{t-1} + \sum_{m=1}^{p} \varphi_m\ \Delta X_{t-m} + \quad \varepsilon_t\ (98)$$

T-statistics are compared with the T-tabulated Dickey Fuller table for series and the Mackinnon table for errors; if the null hypothesis is accepted in one of the models at a given threshold, the process is considered non-stationary, and if the alternative hypothesis is accepted, the process is considered stationary.

II.3.2 Cointegration analysis and Granger causality in the Toda-Yamamoto sense

a) ARDL modeling

AutoRegressive Distributed Lag/ARDL models are dynamic models. They have the particularity of taking into account temporal dynamics (adjustment lag, expectations, etc.) in the explanation of a variable (time series), thus improving forecasts and the effectiveness of policies (decisions, actions, etc.), unlike the simple (non-dynamic) model whose instantaneous explanation (immediate effect or not spread over time) only restores part of the variation in the variable to be explained. Within the family of dynamic models, there are three types of model:

- Autoregressive (AR) models: These are dynamic models in which the lagged dependent variable (its past values) is included among the explanatory variables. The term "autoregressive" refers to the regression of a variable on itself, i.e. on its own lagged values.
- Distributed lag (DL) models: These are dynamic models whose explanatory variables are the initial set of explanatory variables and their lagged or past values. The term "staggered lag" means that the short-term effects of the endogenous variable's explanatory variables are different from the long-term effects. From one point in time to another, the scales of reaction of the endogenous variable to changes in the explanatory variables differ.
- Staggered lag autoregressive (SLAR) models: These models combine the features of two previous models. The explanatory variables include the lagged dependent variable and the past values of the independent variable.

These dynamic models generally suffer from problems of error autocorrelation, with the presence of the lagged endogenous variable as an explanatory variable (AR and ARDL models), and multi-colinearity (DL and ARDL models), which complicates Ordinary Least Squares (OLS) parameter estimation. Here, robust estimation techniques are needed to overcome these problems.

The ARDL model is used for several reasons. Firstly, ARDL takes into account the endogeneity problem by adding lags of dependent and independent variables to the model. Secondly, unlike Johansen's Co-integration, ARDL does not require all variables in the model to integrate at the same order; it can be applied as long as the variables are stationary in level or first differences, or a combination of both. What's more, ARDL applies to a small sample size (Odhiambo, 2009; Zeeshan et al., 2022). Lastly, ARDL can be used to simultaneously estimate the long-term and short-term dynamics of the variables under study. To study the long-term cointegration relationship between variables, the literature provides several econometric techniques. For example, the Engle and Granger (1987) and those of Johansen (1988, 1991) and Pesaran et al. (2001).

Co-integration between series assumes the existence of one or more long-term equilibrium relationships between them, which can be combined with the short-term dynamics of these series in an error-correction (vector) model.

The Co-integration test by Engle and Granger (1987) only helps to check for Co-integration between two integrated series of the same order. It is therefore adapted to the bivariate case and is less effective for multivariate cases. To compensate for this, the Co-integration test of Johansen (1988, 1991) was developed to test co-integration on more than two series, and was designed for multivariate cases. However, while the Johansen test based on vector autoregressive error correction modeling (VECM) is a remedy for the limitations of the Engle and Granger test, it also requires that all series or variables be integrated of the same order. This is not always the case in practice. So, when we have several integrated variables of different orders I(0) and I(1), we can resort to the Co-integration test of Pesaran et al (2001).

We preferred the Pesaran et al (2001) for several reasons. The first is that our series are integrated at different orders I (0) and I (1). The second reason is that it allows us to combine short-term dynamics and long-term effects by referring to the error-correction model. In this research, we use the ARDL model to study the dynamic relationship between CO_2 emissions, consumption of petroleum products (Super and Gasoil), the vehicle fleet, the price of petroleum (Super and Gasoil) and urbanization. The equation is as follows:

$$\Delta \, lnCO_{2t} = \alpha_0 + \sum_{i=1}^{t} \beta_i \, \Delta \, lnCO_{2t-i} + \sum_{i=1}^{t} \rho_i \, \Delta \, lnCPP_{t-i} + \sum_{i=1}^{t} \varphi_i \, \Delta \, lnPAU_{t-i} + \sum_{i=1}^{t} \omega_i \, \Delta \, lnPRX_{t-i} + \sum_{i=1}^{t} \Phi_i \, \Delta \, lnURB_{t-i} + \lambda_1 \, lnCO_{2t-1} + \lambda_2 \, lnCPP_{t-1} + \lambda_3 \, lnPAU_{t-1} + \lambda_4 \, lnPRX_{t-1} + \lambda_5 \, lnURB_{t-1} + \varepsilon_t$$
(99)

With Δ the first difference operator; α_0 the constant; β_i, ρ_i, φ_i, ω_i and Φ_i represent short-term effects; $\lambda_1....\lambda_5$ represent the long-term dynamics of the model and ε_t the error term (white noise); ln CPP represents the logarithm of petroleum product consumption (Super and Gasoil) and ln PRX represents the logarithm of petroleum prices (Super and Gasoil).

Next, to judge Co-integration or the long-term relationship between endogenous and exogenous variables, the Wald restriction test is used. The value of the

F-test is taken by applying the diagnosis of the Wald restriction test coefficient to the parameters of the long-run variable.

The assumptions for Co-integration are :

$H_0 : \lambda_1 = \lambda_2 = \lambda_3 = \lambda_4 = \lambda_5$ (Absence of Co-integration) and,

$H_1 : \lambda_1 \neq \lambda_2 \neq \lambda_3 \neq \lambda_4 \neq \lambda_5$ (Presence of Co-integration)

If the calculated F-stat value is greater than the upper bound value, then the null hypothesis will be rejected, thus concluding the existence of a Co-integration. This means that a long-term relationship exists between the independent and dependent variables. When the F-stat value is below the lower bound value, then the null hypothesis is not rejected; this indicates that the dependent and independent variables have no long-term relationship because they do not show Co-integration. Finally, the result will be considered inconclusive if the F-stat lies between the lower and upper limits.

The Co-integration approach allows us to perform a Fisher test on the selected ARDL equation with appropriate lag lengths. In this study, we used the bounds reported by Narayan (2005) for the small sample size of $30 \leq n \leq 80$ to support the decision on Co-integration estimates. The next step is to estimate long- and short-term results and perform diagnostic tests to ensure model stability. Equation (100) represents short-term models where θi represents the long-term equilibrium adjustment speed after the short-term shock or error correction term.

$$\Delta \, lnCO_{2t} = \beta_0 + \sum_{i=1}^{t} \beta_i \, \Delta \, lnCO2_{t-i} + \sum_{i=1}^{t} \rho_i \, \Delta lnCPP_{t-i} + \sum_{i=1}^{t} \varphi_i \, \Delta lnPAU_{t-i} + \sum_{i=1}^{t} \omega_i \, \Delta \, lnPRX_{t-i} + \sum_{i=1}^{t} \Phi_i \, \Delta \, lnURB_{t-i} + \theta_i ECT_{t-1} + \varepsilon_t$$

$$(100)$$

Checking the model's diagnosis is very important, as it enables the model to be validated. The Breush-Godfrey LM test for serial correlation and the normality test are used to verify serial independence. The ARCH test is also used to check for heteroscedasticity in the model, and the Ramsey Reset test is applied to check for model misspecification. Recursive CUSUM and CUSUM of squares (Brown et al., 1975) are used to detect whether an autoregressive structure is creeping into the model. These tests are used to ensure the stability of model parameters.

b) Granger causality in the sense of Toda-Yamamoto

A number of criticisms of traditional causality tests (principally Granger's) have confirmed the efficacy of Granger's causality test in the sense of "a test of causality". (Toda and Yamamoto, 1995). It should be remembered that the Granger test only applies to stationary series, which makes preliminary tests of series cointegration or analysis of their stationarity essential before verifying any causality between them. However, unit root tests are less effective on small samples and are not always unbiased.

Thus, transforming the series by the first difference, for the sake of stationarity or cointegration, gives good statistical properties while losing information on the level of the series, which level information should not be suppressed as it is enriching in explaining the dynamics of the model under study. It follows that, on small samples, Johansen's cointegration test is sensitive to certain choice parameters which are likely to weaken it: lag (risk of estimating an under-parameterized VAR) and the presence (absence) of deterministic trend in the VAR and/or cointegration space (risk of loss in degrees of freedom).

These parameters create biases that often lead to the rejection of the hypothesis of no cointegration when it is true. This weakness in cointegration results, coupled with the biased nature of unit root tests, reduces the effectiveness of Granger's causality test (random outcome) and pushes (Toda and Yamamoto, 1995) to propose to propose non-sequential procedures for testing causality between series. For these authors, preliminary tests of stationarity and cointegration (Granger's sequential procedures) are of little importance to the economist, who should be more concerned with testing theoretical restrictions (they secure level information). These two authors propose to estimate a level-corrected (over-parameterized) VAR, to be used as a basis for the causality test, under the hypothesis of probable potential cointegration between series, which they integrate into the model without studying it as such (explicitly).

As our variables are integrated at different orders I(0) and I(1), the traditional Granger causality test becomes ineffective. In this case, we resort to the Granger causality test in the sense of Toda and Yamamoto (1995) which is based on a modified Wald test (MWALD). Wolde-Rufael (2005, 2004) and Zapata and Rambaldi (1997) reported that the modified Wald test (MWALD) avoids the problems associated with Granger's ordinary causality test by ignoring any possible non-stationarity or cointegration between series. The approach of Toda and Yamamoto (1995) fits a VAR model to the levels of the variables, thus minimizing the risks associated with any incorrect identification of the order of integration of the series. (Amiri and Ventelou, 2012; Mavrotas and Kelly, 2001). The Granger causality test procedure proposed by Toda-Yamamoto (1995) is as follows:

- Find the order of maximum integration of the series under study d_{max} using stationarity tests;
- Determine the optimal lag of the VAR in sub-study level (k) or autoregressive polynomial (AR) using the information criteria (AIC, SC and HQ);
- Estimating a VAR in augmented order level $k + d_{max}$.

For the estimation of the VAR in increased level, the conditions of stationarity of the series will define the number of lags to be added to the VAR. In fact, for level-stationary series, no lag is added to the VAR (standard test procedure). On the other hand, for I(1) series, a lag is added to the VAR, and so on. The following models are therefore expressed as Granger causality relationships in the Toda-Yamamoto sense.

Model 1: CO emissions$_2$ (CO_2) and supermarket consumption (CSU)

$$lnCO_{2t} = \alpha_0 + \sum_{i=1}^{k} \beta_{1i} \, lnCO_{2t-i} + \sum_{j=k+1}^{d_{max}} \beta_{2j} \, lnCO_{2t-j} +$$
$$\sum_{i=1}^{k} \lambda_{1i} \, ln\, CSU_{t-i} + \sum_{j=k+1}^{d_{max}} \lambda_{2j} \, ln\, CSU_{t-j} +$$
$$\sum_{i=1}^{k} \Phi_{1i} \, ln\, PAU_{t-i} + \sum_{j=k+1}^{d_{max}} \Phi_{2j} \, ln\, PAU_{t-j} +$$
$$\sum_{i=1}^{k} \varphi_{1i} \, ln\, PSU_{t-i} + \sum_{j=k+1}^{d_{max}} \varphi_{2j} \, ln\, PSU_{t-j} +$$
$$\sum_{i=1}^{k} \rho_{1i} \, URB_{t-i} + \sum_{j=k+1}^{d_{max}} \rho_{2j} \, URB_{t-j} + \varepsilon_{1t}$$

$$(101)$$

Model 2: Supermarket consumption (CSU) and CO emissions$_2$ (CO)$_2$

$$lnCSU_t = \alpha_1 + \sum_{i=1}^{k} \beta_{1i} \, lnCO_{2t-i} + \sum_{j=k+1}^{d_{max}} \beta_{2j} \, lnCO_{2t-j} +$$
$$\sum_{i=1}^{k} \lambda_{1i} \, lnCSU_{t-i} + \sum_{j=k+1}^{d_{max}} \lambda_{2j} \, ln\, CSU_{t-j} +$$
$$\sum_{i=1}^{k} \Phi_{1i} \, ln\, PAU_{t-i} + \sum_{j=k+1}^{d_{max}} \Phi_{2j} \, ln\, PAU_{t-j} +$$
$$\sum_{i=1}^{k} \varphi_{1i} \, ln\, PSU_{t-i} + \sum_{j=k+1}^{d_{max}} \varphi_{2j} \, ln\, PSU_{t-j} +$$
$$\sum_{i=1}^{k} \rho_{1i} \, URB_{t-i} + \sum_{j=k+1}^{d_{max}} \rho_{2j} \, URB_{t-j} + \varepsilon_{2t}$$

$$(102)$$

Model 3: CO emissions$_2$ (CO$_2$) and diesel consumption (CGA)

$$lnCO_{2t} = \alpha_0 + \sum_{i=1}^{k} \beta_{1i}\, lnCO_{2t-i} + \sum_{j=k+1}^{d_{max}} \beta_{2j}\, lnCO_{2t-j} +$$
$$\sum_{i=1}^{k} \lambda_{1i}\, ln\, CGA_{t-i} + \sum_{j=k+1}^{d_{max}} \lambda_{2j}\, ln\, CGA_{t-j} +$$
$$\sum_{i=1}^{k} \Phi_{1i}\, ln\, PAU_{t-i} + \sum_{j=k+1}^{d_{max}} \Phi_{2j}\, ln\, PAU_{t-j} +$$
$$\sum_{i=1}^{k} \varphi_{1i}\, ln\, PGA_{t-i} + \sum_{j=k+1}^{d_{max}} \varphi_{2j}\, ln\, PGA_{t-j} +$$
$$\sum_{i=1}^{k} \rho_{1i}\, URB_{t-i} + \sum_{j=k+1}^{d_{max}} \rho_{2j}\, URB_{t-j} + \varepsilon_{1t}$$

(103)

Model 4: Diesel fuel consumption (CGA) and CO emissions$_2$ (CO)$_2$

$$lnCGA_{t} = \alpha_1 + \sum_{i=1}^{k} \beta_{1i}\, lnCO_{2t-i} + \sum_{j=k+1}^{d_{max}} \beta_{2j}\, lnCO_{2t-j} +$$
$$\sum_{i=1}^{k} \lambda_{1i}\, lnCGA_{t-i} + \sum_{j=k+1}^{d_{max}} \lambda_{2j}\, ln\, CGA_{t-j} +$$
$$\sum_{i=1}^{k} \Phi_{1i}\, ln\, PAU_{t-i} + \sum_{j=k+1}^{d_{max}} \Phi_{2j}\, ln\, PAU_{t-j} +$$
$$\sum_{i=1}^{k} \varphi_{1i}\, ln\, PGA_{t-i} + \sum_{j=k+1}^{d_{max}} \varphi_{2j}\, ln\, PGA_{t-j} +$$
$$\sum_{i=1}^{k} \rho_{1i}\, URB_{t-i} + \sum_{j=k+1}^{d_{max}} \rho_{2j}\, URB_{t-j} + \varepsilon_{2t}$$

(104)

CO emissions$_2$ cause super and diesel consumption if $\lambda_{fi} \neq 0$, $\forall i = 1,2,\ldots,k$ in equation (101) and equation (103) respectively. Similarly, super and diesel consumption cause emissions of CO_2 si $\beta_{1i} \neq 0$, $\forall i = 1,2,\ldots,k$ respectively in equation (102) and equation (104). There is bidirectional causality between emissions of CO_2 emissions and super and diesel consumption if $\lambda_{fi} \neq 0$ and $\beta_{1i} \neq 0$, $\forall i = 1,2,\ldots,k$ in equations (101), (102), (103) and (104) respectively. Finally, there is no causality between emissions of CO_2 emissions and super and diesel consumption if $\lambda_{fi} = \beta_{1i} = 0$, $\forall i = 1,2,\ldots,k$ in equations (101), (102), (103) and (104) respectively. The same causal inference can be made with the other models. We could include many more models by making each of the variables (in turn), the subject of their respective equations. While this is possible, the results are not always interesting, or worse still, they are not interpretable. So we model and test only those equations that are interesting and interpretable.

II.4 THE Forecasts

II.4.1 Use of multiple linear regression models

In order to study annual CO_2 emissions for future years, a multiple linear regression model can be used. We specify the emissions of CO_2 (CO_{2t}) as a function of super and diesel fuel consumption (CSU_t and CGA_t), vehicle fleet (PAU_t), super and diesel prices (PSU_t and PGA_t), urbanization (URB_t) and u_t an error term. The dynamic multiple linear function is written in the following form:

CO emissions model$_2$ and supermarket fuel consumption

$$CO_{2t} = \beta_0 + \beta_1 CSU_t + \beta_2 PAU_t + \beta_3 PSU_t + \beta_4 URB_t + \beta_5 CO_{2t-2} + \beta_6 CSU_{t-3} + u_t$$

$$(105)$$

Model of CO emissions$_2$ and diesel consumption

$$CO_{2t} = \beta_0 + \beta_1 CGA_t + \beta_2 PAU_t + \beta_3 PGA_t + \beta_4 URB_t + \beta_5 CO_{2t-2} + \beta_6 CGA_{t-3} + u_t$$

$$(106)$$

All variables are in logarithmic form. $\beta_i; i = 0, 1, 2, 3, 4, 5, 6$ are the regression coefficients. $(t - i)$ denotes lagged variables. It is expected that β_1 and β_6 are expected to be negative, while β_2, β_3, β_4 and β_5 are expected to be positive for the usual economic reasons.

When none of the variables in equations (23) and (24) are level-stationary, if they are allowed in their level form, it becomes impossible to perform conventional inference on the coefficients, as this leads to spurious regression. An alternative is to convert the data to make them stationary, thus obtaining consistent forecasting equations. As we saw earlier, a good solution is to differentiate the data to make them stationary. The resulting equation will therefore involve the first derivative of the variables. Thus, the new model (for each) takes the following form:

$$\Delta CO_{2t} = \beta_0 + \beta_1 \Delta CSU_t + \beta_2 \Delta PAU_t + \beta_3 PSU_t + \beta_4 \Delta URB_t + \beta_5 \Delta CO_{2t-2} + \beta_6 \Delta CSU_{t-3} + u_t$$

$$(107)$$

$$\Delta CO_{2t} = \beta_0 + \beta_1 \Delta CGA_t + \beta_2 \Delta PAU_t + \beta_3 \Delta PGA_t + \beta_4 \Delta URB_t + \beta_5 \Delta CO_{2t-2} + \beta_6 \Delta CGA_{t-3} + u_t$$

$$(108)$$

Where Δ is the percentage difference operator and the variables are always defined as in equations (110) and (111). Data from the period 2008 to 2019 are used to estimate the model coefficients. The results of the CO_2 are presented in the next chapter.

II.4.2 Residual tests, precision and validation

The Breusch-Godfrey LM test for serial correlation is applied to equations (107) and (108), indicating the absence of serial correlation in the residuals. Heteroskedasticity is tested using the Breusch-Pagan-Godfrey test.

An accuracy measure is defined as the difference between the actual value and the predicted value. There are a large number of performance measures in the literature, each with its own advantages and limitations. The most common are: root mean square error (RMSE), mean absolute error (MAE) and mean absolute percentage error (MAPE). These quantities are defined as follows:

$$RMSE = \sqrt{\frac{1}{N}\sum_{t=1}^{N}(CO_{2t} - \widehat{CO_{2t}})^2}$$

(109)

$$MAE = \frac{1}{N}\sum_{t=1}^{N}\left|CO_{2t} - \widehat{CO}_{2t}\right|^2$$

(110)

$$MAPE = \frac{1}{N}\sum_{t=1}^{N}\left|\frac{CO_{2t}-\widehat{CO}_{2t}}{D_t}\right|.100$$

(111)

Where CO_{2t} represents actual CO emissions$_2$, $\widehat{CO}_{2t}$ the CO emissions$_2$ predicted by the model and N is the number of observations.

To assess the model's performance, an error analysis based on RMSE, MAE and MAPE is provided. In addition, to give a full indication of the validity of the model proposed in this paper, a graphical representation of the fit between the actual consumption profile and the sample regression demand function is provided.

A further validation test is carried out by estimating the forecasting equation on data from 2008 to 2015, so that the remaining 3 years (i.e. 2015 to 2018) are reserved for evaluating the model on new data. In this way, it is possible to assess the validity of the model on real data. Actual and predicted demands are presented in the next chapter. In addition, a regression model is appropriate when the residual plots are randomly dispersed around the horizontal axis.

Thus, on the basis of the validation tests carried out, it will be possible to say that the model represented by equations (110) and (111) is a valid model for estimating demand for petroleum products in Cameroon if all performance measures are guaranteed.

II.4.3   Forecasting emissions of CO_2 emissions using multiple linear regression models

To generate future emission values for CO_2 we need to obtain new values for $CSU_t, CGA_t, PAU_t, PSU_t, PGA_t$ and URB_t insert them into Eqs. (110) and (111), and generate CO_{2t} for $t > 2019$. The independent variables are estimated and predicted by simple linear regression over time. They can therefore be represented by the following equations:

$$CSU_t = a_1 + b_1 t \tag{112}$$

$$CGA_t = a_2 + b_2 t \tag{113}$$

$$PAU_t = a_3 + b_3 t \tag{114}$$

$$PSU_t = a_4 + b_4 t \tag{115}$$

$$PGA_t = a_5 + b_5 t \tag{116}$$

$$URB_t = a_6 + b_6 t \tag{117}$$

Where $a_1, a_2, a_3, a_4, a_5, a_6, b_1, b_2, b_3, b_4, b_5, b_6$ are regression coefficients and t is the time parameter indicating the year. Assuming that 2018 is the current year, equations (112-117) are used to forecast real super and diesel consumption, the vehicle fleet, real super and diesel prices and real urbanization for emissions of CO_2 emissions over the forecast period. The estimated coefficients of these equations are presented in the next chapter.

By inserting the values of the explanatory variables obtained from the trend regression in equations (112-117) into equations (107) and (108), we can forecast emissions of CO_2 emissions for the following period of years.

II.4.4 THE Grey's multivariable sequential convolution model optimized by genetic algorithms (GMC(1, N) - GA)

The GMC (1, N) model is based on six stages (Shen et al., 2021; Tien, 2012):

Step 1: Construction of input sequences

Let $X_1^{(0)}, X_2^{(0)}, ..., X_n^{(0)}$ and the variables of a grey system. $X_1^{(0)}, X_2^{(0)}, ..., X_n^{(0)}$ are used to construct the input sequence. Each variable in the input sequence is defined by Eq. (118) :

$$X_i^{(0)} = \left\{ x_i^{(0)}(1), x_i^{(0)}(2), ..., x_i^{(0)}(k) \right\}; \ i = 1, 2, ..., n; \ k \geq 4$$

(118)

$X_i^{(0)}$ is a non-negative sequence; k is the number of samples of the $i^{ème}$ input variable, and the exponent (0) denotes the original sequences. $X_1^{(0)}$ is the output and $X_i^{(0)}, i = 2, 3, ..., n$ are the variables.

Step 2: Accumulated generation operation

When $X_i^{(0)}$ is subjected to the accumulated generation operation, we obtain the following Eq. (119) :

$$X_i^{(1)} = \left\{ x_i^{(1)}(1), x_i^{(1)}(2), ..., x_i^{(1)}(k) \right\}$$

(119)

Where:

$$x_i^{(1)}(t) = \sum_{m=1}^{t} x_i^{(0)}(m); \quad t = 2, 3, \dots, T$$

(120)

The exponent (1) in Eq. (120) represents the accumulated first-order generation of the original sequences. $x_i^{(1)}(k)$ increases continuously. The sequence is therefore monotonic (Hamzacebi and Es, 2014b).

Step 3: Designing the system's background value

The system background value is defined as follows in Eq. (121):

$$Z_1^{(1)}(t) = \theta X_1^{(1)}(t) + (1 - \theta) X_1^{(1)}(1 - t); \quad t = 2, 3, \dots, T$$

(121)

The horizontal adjustment coefficient θ is taken as $0 < \theta < 1$ (Ding et al., 2018; Hamzacebi and Es, 2014b). The value of θ is generally taken as 0.5, but it should be chosen in such a way as to reduce prediction errors (Hamzacebi and Es, 2014b).

Step 4: Establish Grey's systems of equations

Grey's system is formed by linking known and unknown sequences:

$$dx_1^{(1)}(t)/dt + ax_1^{(1)}(t) = b_2 x_2^{(1)}(t) + b_3 x_3^{(1)}(t) + \cdots + b_n x_n^{(1)}(t) + u$$

(122)

a is the development coefficient, u is the GMC(1,n) parameter, while $b_{j=1,\dots,n}$ are Grey's input coefficients. Tien (2012) begins by considering the right-hand side of Eq. (122) in terms of $f(t)$. Then, using the trapezoid formula and integrating the two sides of Eq. (122) from $t - 1$ à tEq. (122) is approximated by the following difference equation (Tien, 2012):

$$x_1^{(0)}(t) + az_1^{(1)}(t) = b_2 z_2^{(1)}(t) + b_3 z_3^{(1)}(t) + \cdots + b_n z_n^{(1)}(t) + u \qquad (123)$$

Consequently, Eq. (123) can be considered as a system of linear equations in relation to the coefficients $[a\ b_2\ b_3\ ...\ b_n\ u]^T$. These coefficients are calculated by the method of least **squares** (Bahrami et al., 2014; Deng, 1982; Ding et al., 2018; Hamzacebi and Es, 2014b). Thus, applying the least-squares method, $[a\ b_2\ b_3\ ...\ b_n\ u]^T$ are determined as shown in Eq. (124) :

$$A = \begin{bmatrix} a \\ b_2 \\ b_3 \\ \vdots \\ b_n \\ u \end{bmatrix} = (B^T B)^{-1} B^T Y$$

$$(124)$$

Where
$$B = \begin{bmatrix} -z_1^{(1)}(2) & x_2^{(1)}(2) & x_3^{(1)}(2) & ... & x_n^{(1)}(2) \\ -z_1^{(1)}(3) & x_2^{(1)}(3) & x_3^{(1)}(3) & \cdots & x_n^{(1)}(3) \\ \vdots & \vdots & \vdots & \ddots & \vdots \\ -z_1^{(1)}(n) & x_2^{(1)}(n) & x_3^{(1)}(n) & & x_n^{(1)}(m) \end{bmatrix}; \quad Y = \begin{bmatrix} x_1^{(0)}(2) \\ x_1^{(0)}(3) \\ \vdots \\ x_1^{(0)}(n) \end{bmatrix}$$

Step 5: Solve the system's differential equation

The solution of Eq. (122) is (Tien, 2012) :

$$\hat{x}_1^{(1)}(t) = x_1^{(0)}(1)e^{a(1-t)} + \int_1^t e^{a(\tau-t)} f(\tau)d\tau; \quad t \geq 2 \qquad (125)$$

The value of which can only be approximated numerically, due to the presence of a convolution integral. With the trapezoid formula, Eq. (125) becomes :

$$\hat{x}_1^{(1)}(t) = x_1^{(0)}(1)e^{a(1-t)} + 0.5h(t)\sum_{i=2}^{n}\left[f(\tau)e^{a(\tau-t)} + f(\tau - 1)e^{a(\tau-t-1)}\right]; \ t \geq 2 \qquad (126)$$

Recall that $\hat{x}_1^{(1)}(1) = x_1^{(0)}(1)$and $h(t)$ in Eq. (126) are defined as follows: $h(t) = 0$ if $t < 2$on the other hand $h(t) = 1$

Step 6: Reverse the accumulated generation operation

Finally, the predicted values of $\hat{x}_1^{(0)}(t)$ are obtained by the accumulated generation operation as shown in Eq. (127) :

$$\hat{x}_1^{(0)}(t) = \hat{x}_1^{(1)}(t) - \hat{x}_1^{(1)}(t-1); \qquad t \geq 2$$
$$(127)$$

GMC (1, N) is a significant improvement on the basic GM (1, N) version in terms of quality. However, it has several obvious shortcomings, as demonstrated by Shen (Shen et al., 2021) has demonstrated. GMs accuracy is affected by k, θ, a and $b_{j=1,...,n}$ (Ding et al., 2017; Hamzacebi and Es, 2014b; Tseng et al., 2001).. In this study, the best values of these parameters are determined experimentally (instead of using arbitrary values reported in the literature). Thus, a new GA-optimized GMC (1, N) sequential model is proposed to address these shortcomings in order to improve accuracy, stability and CPU runtime performance.

k, θ, a and $b_{j=1,...,n}$ are found in the literature at 0.5 and 4 for θ and k respectively, while a and $b_{j=1,...,n}$ are calculated only once during the modeling phase (Hamzacebi and Es, 2014a). However, periodic values of these parameters can significantly increase the accuracy of GMC (1, N) (Hamzacebi and Es, 2014a). Thus, instead of using values reported in the literature, we instead calculate experimentally (using AC- GA) for each forecast period h.

The sequential mechanism uses the most recent simulations for prediction by describing the latest characteristics of the data. This enables the sequential mechanism to improve forecast accuracy. This mechanism uses p sequences to model GMC (1, N) and q consumption data for forecasting. At each loop, new values of k, θ, a and $b_{j=1,\ldots,N}$ are calculated and optimized by GA. The process is described below:

i.   Construct GMC(1,n) using $X_i^{(0)}(1), X_i^{(0)}(2), \ldots, X_i^{(0)}(p)$ and calculate k, θ, a and $b_{j=1,\ldots,n}$

ii.   Optimize k, θ, a and $b_{j=1,\ldots,n}$ with AC-GA and forecasts $\hat{X}_1^{(0)}(p+1), \hat{X}_1^{(0)}(p+2), \ldots, \hat{X}_i^{(0)}(p+q)$

iii.   Remove $X_i^{(0)}(1), X_i^{(0)}(2), \ldots, X_i^{(0)}(p)$ from the sequence, reconstruct GMC(1,n) with the most recent p sequences, i.e. $X_i^{(0)}(q+1), X_i^{(0)}(q+2), \ldots, et\ calculer\ X_i^{(0)}(q+p)$ new values of k, θ, a and $b_{j=1,\ldots,n}$

iv.   Use AC-GA to optimize new calculations k, θ, a and $b_{j=1,\ldots,n}$ and forecasting $\hat{X}_1^{(0)}(p+q+1), \hat{X}_1^{(0)}(p+q+2), \ldots, \hat{X}_i^{(0)}(p+2q)$

v.   Repeat steps (ii) to (iv) until all the required CO2 emissions data have been predicted.

This process is called GA-optimized GMC (1, N) sequential prediction (Sequential GMC (1, N) - GA). The above instructions have been written with CA (steps ii and iv). This is a GMC (1, N) - GA sequential with AC. By editing the "with AC" instruction in steps ii and iv, it becomes a GMC (1, N) - GA sequential without AC.

Conclusion

This chapter details the method used in this study. It involves testing the stationarity of the variables under study and determining the number of lags and the information criterion to be used. Co-integration analysis enables us to identify the long-term and short-term effects of our ARDL model. The Granger causality test in the Toda-Yamamoto sense enables us to establish the precise influences between CO2 emissions, the consumption of petroleum products (Super and Gasoil), the vehicle fleet, the price of petroleum products (Super and Gasoil) and urbanization. Finally, the GMC (1, N) - GA sequential forecasting model is used to assess Chad's future oil consumption. CO_2 future consumption in Chad. The following chapter gives a detailed presentation of the results obtained and highlights a series of discussions to better understand them.

CHAPTER III
SIMULATED RESULTS, ANALYSIS AND DISCUSSION

Introduction

In this chapter, we discuss the results of the functional link between emissions of CO_2 and its regressors (consumption of petroleum products, price of petroleum products at the pump, vehicle fleet, urbanization and white noise, which must be identically and normally distributed). In particular, we present all the usual tests to prove that a model is suitable for use in forecasting. At the same time, we quantify their short- and long-term effects, as well as existing causalities between variables. Once the model has been validated (using the usual validation criteria presented in the previous chapter), we estimate future CO_2 emissions, based on the established Grey model. Based on the results obtained, we conclude this chapter with policy recommendations for decision-makers in Chad. This will guide them to better implement the strategy related to emissions of CO_2.

III.1 Simulation of results and analysis

III.1. 1 Descriptive statistics and correlation between variables

a) Descriptive statistics

Table 6 shows that standard deviations are low overall for the different series. This observation can be explained by the logarithmic transformation of our series, which has the induced effect of attenuating the variances between variable values. Also, the different variables under study are normally distributed with regard to the Jarque-Bera probability (Udemba and Yalçıntaş, 2021)..

Table 6 Descriptive statistics of variables

	$\ln CO_2$	ln CSU	ln CGA	ln PSU	ln PGA	ln PAU	ln URB
Average	7.382964	18.61211	19.35107	6.260847	6.365421	9.998253	1.313470
Median	7.682440	19.15376	19.52802	6.232292	6.313750	9.821792	1.334660

Maximum	7.779049	19.34589	19.72302	6.476972	6.620073	10.60142	1.360722
Minimum	6.291569	17.20753	18.43153	5.940171	6.152733	9.675897	1.241889
Standard deviation	0.494560	0.883419	0.401245	0.157119	0.170029	0.360592	0.045469
Asymmetry coefficient	-1.157113	-0.662033	-1.020883	-0.117847	0.540024	0.795804	-0.520180
Flattening coefficient	2.961098	1.594397	3.142207	2.815933	2.011278	2.003185	1.612036
Jarque-Bera	2.678577	1.864434	2.094515	0.044716	1.072038	1.763427	1.504396
Probability	0.262032	0.393680	0.350899	0.977890	0.585073	0.414073	0.471329
Sum	88.59556	223.3454	232.2128	75.13017	76.38505	119.9790	15.76164
Sum of squares	2.690488	8.584712	1.770977	0.271550	0.318007	1.430289	0.022742
Comments	12	12	12	12	12	12	12

Note. ln CO_2 Emissions of CO_2 emissions in logarithm; **ln CSU**: Consumption of super in logarithm; **ln CGA**: Consumption of diesel in logarithm; **ln PSU**: Price of super in logarithm; **ln PGA**: Price of diesel in logarithm; **ln PAU**: Car fleet in logarithm; **ln URB**: Urbanization in logarithm.

b) Correlation between variables in model 1

Table 7 shows the correlation structure between the variables in Model 1. The correlation coefficient between CO_2 emissions and supermarket consumption is 0.95. This coefficient is positive and fairly strong (degree of association greater than 0.5). The same applies to CO_2 emissions and urbanization, for which the correlation coefficient is 0.85. As a result, supermarket consumption is positively and strongly correlated with CO_2 à 0.95%. Thus, an upward trend in supermarket consumption is accompanied by an increase in emissions of CO_2 emissions by 0.95%. Similarly, urbanization is positively and strongly correlated with emissions of CO_2 à 0.85%.

Thus, an upward trend in supermarket consumption is accompanied by an increase in emissions of CO_2 emissions by 0.95%. On the other hand, the price of super and the vehicle fleet are negatively correlated with emissions of CO_2 emissions by 0.48% and 0.38% respectively. Thus, an increase in the price of super and the number of cars on the road are accompanied by a decrease in emissions of CO_2 emissions by 0.48% and 0.38% respectively.

Table 7 Correlation of variables in model 1

	ln CO_2	ln CSU	ln PSU	ln PAU	ln URB
ln CO_2	1				
ln CSU	0.95	1			
ln PSU	-0.48	-0.44	1		
ln PAU	-0.38	-0.25	0.26	1	
ln URB	0.85	0.94	-0.40	-0.30	1

Note. **lnCO_2** Emissions of CO_2 emissions in logarithm; **lnCSU**: Supermarket consumption in logarithm; **lnPSU**: Supermarket price in logarithm; **lnPAU**: Car fleet in logarithm; **lnURB**: Urbanization in logarithm.

c) Correlation between variables in model 2

Table 8 shows the correlation structure between the variables in Model 2. The correlation coefficient between CO_2 emissions and diesel consumption is 0.87. This coefficient is positive and fairly strong (degree of association greater than 0.5). The same applies to CO_2 emissions with urbanization, for which the correlation coefficient is 0.85. As a result, diesel consumption is positively and strongly correlated with emissions of CO_2 à 0.87%. Thus, an upward trend in diesel consumption is accompanied by an increase in emissions of CO_2 emissions by 0.87%. Similarly, urbanization is positively and strongly correlated with emissions of CO_2 à 0.85%.

Thus, an upward trend in diesel consumption is accompanied by an increase in emissions of CO_2 emissions by 0.87%. On the other hand, the price of diesel and the vehicle fleet are negatively correlated with emissions of CO_2 emissions by 0.76% and 0.38% respectively. Thus, an increase in the price of diesel and the number of vehicles on the road are accompanied by a fall in emissions of CO_2 emissions by 0.76% and 0.38% respectively.

Table 8 Correlation of variables in model 2

	$\ln CO_2$	lnCGA	lnPGA	lnPAU	lnURB
ln CO_2	1				
ln CGA	0.87	1			
ln PGA	-0.76	-0.80	1		
ln PAU	-0.38	-0.21	0.05	1	
ln URB	0.85	0.76	-0.74	-0.30	1

Note. **lnCO_2** Emissions of CO_2 emissions in logarithm; **lnCGA**: Diesel consumption in logarithm; **lnPGA**: Diesel price in logarithm; **lnPAU**: Car fleet in logarithm; **lnURB**: Urbanization in logarithm.

III.1.2 Variable stationarity and optimal selection criterion

a) Variable stationarity

The results of the Augmented Dickey-Fuller (ADF) and Phillips-Perron (PP) unit root tests contained in Table 9 confirm that emissions of CO_2 emissions and diesel consumption are stationary at level I (0). Supermarket consumption, supermarket prices, diesel prices and the number of cars on the road are stationary at level I (1). Finally, urbanization is stationary either at level I (0) with the ADF test, or at first difference I (1) with the PP test. In general, none of the variables is stationary at second difference I (2), which satisfies the requirements of the ARDL boundary co-integration test.

Table 9 Stationarity of variables

Variables	Level		Difference 1 ère		Decisions
	ADF	PP	ADF	PP	
ln CO_2	-3.38** (0.03)	-4.50*** (0.00)	—	—	I(0)
ln CSU	1.62 (0.96)	1.40 (0.94)	-1.81** (0.04)	-1.81** (0.04)	I(1)
ln CGA	-3.01** (0.04)	-3.01 (0.04)	—	—	I(0)
ln PSU	-0.42 (0.50)	-0.68 (0.39)	-4.36*** (0.00)	-4.49*** (0.00)	I(1)
ln PGA	-0.94 (0.28)	-0.65 (0.41)	-4.67*** (0.00)	-4.64*** (0.00)	I(1)
ln PAU	-0.76 (0.36)	-1.64 (0.11)	-3.08*** (0.00)	-4.31*** (0.00)	I(1)
ln URB	-3.53** (0.03)	1.79 (0.97)	—	-1.96 (0.04)	I(0) or I(1)

Note: ln CO_2 : Emissions of CO_2 in logarithm; **ln CSU**: Consumption of super in logarithm; **ln CGA**: Consumption of diesel in logarithm; **ln PSU**: Price of super in logarithm; **ln PGA**: Price of diesel in logarithm; **ln PAU**: Car fleet in logarithm; **ln URB**: Urbanization in logarithm; ***Significance at 1%; **Significance at 5%; (.) : Probability

b) Optimal selection criteria

i. Optimal selection criterion for model 1

Table 10 shows that all the information criteria (AIC, SIC and HQ) converge towards a number of delays of 1 for model 1.

Table 10 Selection of the number of delays for model 1

Lag	LogL	LR	FPE	AIC	SC	HQ
0	38.65636	NA	1.52e-09	-6.119338	-5.938477	-6.233346
1	135.9313	88.43174*	5.77e-15*	-19.260233*	-18.175066*	-19.944428*

Note:
* indicates the order of delays selected by the criterion
LR: serially modified test statistic (each test at the 5% level)
FPE: final prediction error
AIC: Akaike information criterion
SC: Schwarz information criterion
HQ: Hannan-Quinn information criterion

ii. Optimal selection criterion for model 2

Table 11 shows that all the information criteria (AIC, SIC and HQ) converge towards a number of delays of 1 for model 2.

Table 11 Selection of the number of delays for model 2

Lag	LogL	LR	FPE	AIC	SC	HQ
0	42.10353	NA	8.11e-10	-6.746097	-6.565235	-6.860104
1	155.6468	103.2212*	1.60e-16*	-22.84487*	-21.75971*	-23.52892*

Note:
* indicates the order of delays selected by the
criterion
LR: serially modified test statistic (each test at the 5% level)
FPE: final prediction error

AIC: Akaike information criterion
SC: Schwarz information criterion
HQ: Hannan-Quinn information criterion

Overall, models 1 and 2 have a delay number of 1.

III.1.3 ARDL model estimation and long- and short-term dynamics

a) ARDL model estimation and long- and short-term dynamics of model 1

i.   ARDL model estimation for model 1

Table 12 shows the ARDL model estimate for model 1. Thus, the estimated model (1,1,1,1,0) is globally good and explains 98% of the dynamics of CO_2 emissions in Chad from 2008 to 2019.

Table 12ARDL model estimation for model 1

ARDL(1,1,1,1,0) model				
Variable	Coefficient	Std. Error	t-Statistic	Prob.*
ln CO_2(-1)	1.530343	0.250280	6.114533	0.0088***
ln CSU	-0.044877	0.183946	-0.243968	0.8230
ln CSU(-1)	-0.502506	0.094683	-5.307245	0.0131**
ln PSU	1.326800	0.241682	5.489869	0.0119**
ln PSU(-1)	-0.908701	0.257737	-3.525690	0.0388**
ln PAU	0.116668	0.083449	1.398072	0.2565
ln PAU(-1)	0.077904	0.044183	1.763240	0.1761
ln URB	1.414361	2.295467	0.616154	0.5814

R² : 0.996824
Adjusted R²: 0.989413
Note: ***Significativity at 1%; **Significativity at 5%.

ii.   Graphical values of the SIC information criterion for model 1

Fig. 14 plots the smallest value of the SIC information criterion that offers the fewest delays. It thus confirms that the ARDL (1,1,1,1,0) model is the most optimal of the others.

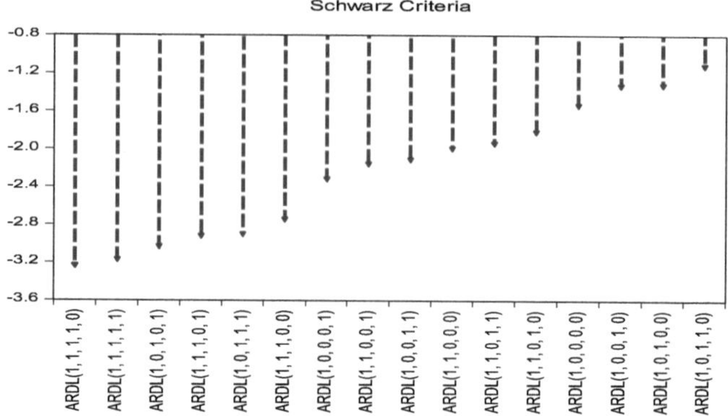

Figure 14SIC graphical representation for Model 1

iii.   Model 1 diagnostic test results

Table 13 shows the results of the robustness tests for model 1. With regard to these tests, which help diagnose the estimated ARDL (1,1,1,1,0) model, we note a good specification of this model, an absence of error autocorrelation, an absence of heteroscedasticity, and errors that follow a normal distribution. The null hypothesis is accepted for all these tests (P-value > 5%). Our model is thus statistically validated

Table 13 Robustness test results for model 1

Robustness tests	P-value	Results
Ramsey Reset test (Stability test)	0.67	The model is perfectly specified
Breusch-Godfrey Serial Correlation LM Test	0.46 0.86	No serial correlation
Jarque-Bera (Normality test)	0.44	Residues are normally distributed
ARCH Test		

(Heteroscedastici ty test)	No evidence of heteroscedasticit y

In addition to these various tests, we also perform a test of the model's graphical stability, namely the cumulative sum of squares of the recursive residuals (CUSUM AT SQUARE). The graphical representation of CUSUM AT SQUARE is given in Fig. 15. According to the guideline, if the plots remain within the 5% critical limit, this means that the model parameters are stable and consistent. The model plot shows that CUSUM AT SQUARE is within limits and over time for the Chad data.

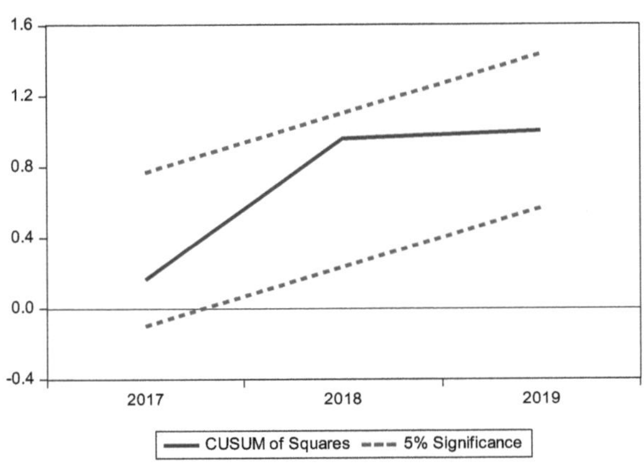

Figure 15 Graphical representation of CUSUM of squares for model 1

iv. Co-integration test results

Table 14 shows the Co-integration results for model 1. The results of the Co-integration test at the bounds confirm the existence of a Co-integration relationship between the series studied (the value of F-stat is > that of the upper bound); this makes it possible to estimate the long-run effects of supermarket consumption, supermarket price, car fleet and urbanization on CO_2

Table 14 Co-integration results for model 1

Test Statistic	Value	Significance	I(0)	I(1)
F-statistics	27.39011	10%	1.9	3.01
K	4	5%	2.26	3.48
		2.5%	2.62	3.9
		1%	3.07	4.44

v.   Long-term and short-term dynamics

Table 15 shows the long-term results. The linear impact of supermarket consumption is positive and significant at the 5% level. Thus, a 1% increase in supermarket consumption would lead to a 1.03% increase in CO emissions$_2$. In addition, the linear impact of the vehicle fleet is positive and significant at the 10% level. Thus, a 1% increase in the number of cars on the road would lead to a 0.36% increase in CO_2 emissions. The price of super and urbanization remain insignificant.

Table 15 Long-term results for model 1

Variable	Coefficient	Std. Error	t-Statistic	Prob.
LCSU	1.032130	0.255539	4.039035	0.0273**
LPSU	-0.788356	0.409104	-1.927031	0.1496
LPAU	0.366880	0.123391	-2.973317	0.0589*
LURB	-2.666882	3.881973	-0.686991	0.5414

Note: **Significativity at 5%; *Significativity at 10%.

Table 16 shows the short-term results. The error correction term (ECT) is negative and significant; this allows us to say that the error correction model is valid and that emissions of CO_2 emissions are approaching their long-term equilibrium level with an adjustment speed of 53.03%, thanks to the contribution of supermarket consumption, the supermarket price and the vehicle fleet. In fact, supermarket consumption is insignificant. Whereas the price of super and the number of cars on the road would have a linear impact of 1% and 5% respectively. Thus, a 1% increase in the price of super and in the number of cars on the road would lead to an increase in emissions of CO_2 emissions by 1.32% and 0.11% respectively.

Table 16 Short-term results for model 1

Variable	Coefficient	Std. Error	t-Statistic	Prob.
D(LCSU)	-0.044877	0.035809	-1.253213	0.2989
D(LPSU)	1.326800	0.071525	18.550060	0.0003***
D(LPAU)	0.116668	0.026959	4.327613	0.0227**
ECT(-1)	-0.530343	0.029668	17.876000	0.0004***

Note: ***Significativity at 1%; **Significativity at 5%; *Significativity at 10%.

vi. Causality between variables in model 1

After estimating the long-term and short-term results, we performed a Granger causality test in the Toda-Yamamoto sense. The results of the Granger causality test in the Toda-Yamamoto sense are shown in Table 17. In this table, two bidirectional causalities emerge: a bidirectional causality between the consumption of super and the price of super, and a bidirectional causality between emissions of CO_2 and the price of super; and five unidirectional causalities: a unidirectional causality from super consumption to emissions of CO_2emissions, a unidirectional causality from CO_2 emissions to the car fleet, a unidirectional causality from CO_2 emissions to urbanization, unidirectional causality from supermarket consumption to vehicle fleet, and unidirectional causality from urbanization to vehicle fleet. Fig. 16 shows the results of Granger causality in the Toda-Yamamoto sense for Model 1.

Table 17 Granger causality results in the Toda-Yamamoto sense for model 1

Dependent variables	Explanatory variables					Decisions
	ln CO_2	ln CSU	ln PSU	ln PAU	ln URB	
ln CO_2		5.71***	9.89***	0.75	0.22	CSU → CO_2 PSU → CO_2
ln CSU	0.02		28.30***	0.63	0.02	PSU → CSU
ln PSU	11.44***	3.63*		2.09	0.00	CO_2 → PSU CSU → PSU
ln PAU	3.16*	5.14**	0.22		6.07**	CO_2 → PAU, CSU → PAU URB → PAU
ln URB	12.48***	2.11	10.75***	0.04		CO_2 → URB PSU → URB

Note: (.) : Probability (P-value); ***Significativity at 1%; **Significativity at 5%; *Significativity at 10%.

The → symbol represents unidirectional causality

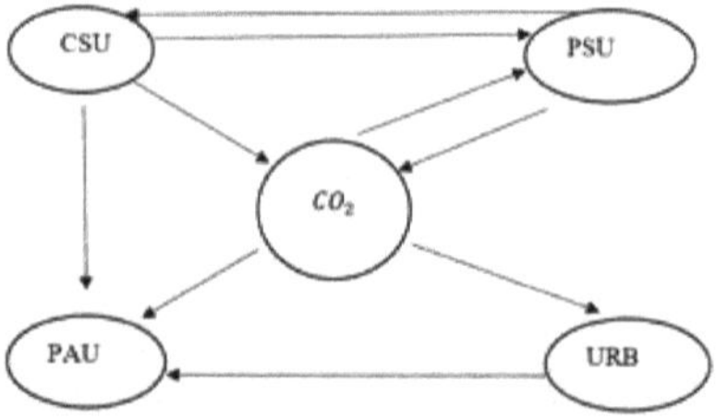

Figure 16 Graphical summary of the Toda-Yamamoto causality test for model 1

b) Estimation of the ARDL model and long- and short-term dynamics of model 2

i. ARDL estimation of model 2

Table 18 shows the ARDL model estimate for model 2. Thus, the estimated model (1,1,1,1,1) is globally good and explains 98% of the dynamics of CO_2 emissions in Chad from 2008 to 2019.

Table 18ARDL model estimation for model 2

ARDL(1, 1, 1, 1, 1)				
Variables	Coefficient	Std. Error	t-Statistic	Prob.*
LCO_2 (-1)	1.196371	0.604495	1.979125	0.1864
LCGA	0.102556	0.472682	0.216966	0.8484
LCGA(-1)	-0.363783	0.237313	-1.532924	0.2650
LPGA	2.083778	0.620590	3.357734	0.0784
LPGA(-1)	-1.178990	0.330604	-3.566165	0.0704
LPAU	0.245617	0.151936	1.616588	0.2474
LPAU(-1)	-0.145649	0.152750	-0.953513	0.4410
LURB	5.415729	2.177097	2.487592	0.1307
LURB(-1)	-7.715484	1.740924	-4.431833	0.0473
R²	0.996872			
Adjusted R²	0.984359			

Note: ***Significativity at 1%; **Significativity at 5%.

ii. Graphical values for the SIC information criterion in Model 2

Fig. 17 plots the smallest value of the SIC information criterion that offers the fewest delays. It confirms that the ARDL (1,1,1,1,0) model is the most optimal of the others.

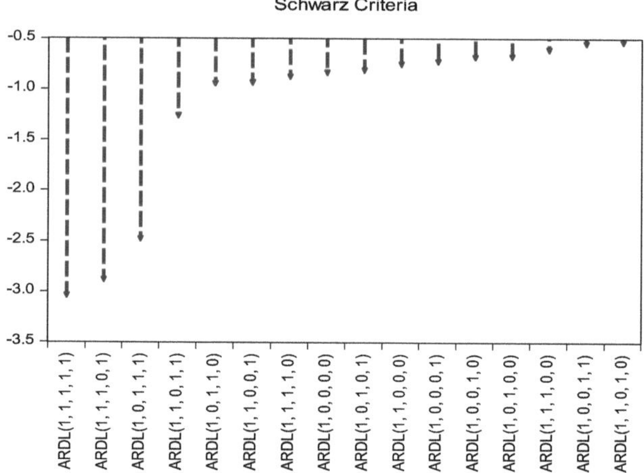

Figure 17SIC graphical representation of model 2

iii.   Model 2 diagnostic test results

Table 19 shows the results of the robustness tests for model 1. With regard to these tests, which help diagnose the estimated ARDL (1,1,1,1,0) model, we note a good specification of this model, an absence of error autocorrelation, an absence of heteroscedasticity, and errors that follow a normal distribution. The null hypothesis is accepted for all these tests (P-value > 5%). Our model is thus statistically validated

Table 19 Results of model 2 robustness tests

Robustness tests	P-value	Results
Ramsey Reset test (Stability test)	0.057	The model is perfectly specified
Breusch-Godfrey Serial Correlation LM Test	0.49	No serial correlation
	0.064	
Jarque-Bera (Normality test)		Residues are normally distributed
	0.78	
ARCH Test		

(Heteroscedasticity test)	No evidence of heteroscedasticity

In addition to these various tests, there are also tests of the model's graphical stability, namely the cumulative sum (CUSUM) and the cumulative sum of squares of the recursive residuals (CUSUM SQUARE). The graphical representation of CUSUM and CUSUM AT SQUARE is shown in Figure 18. According to the guideline, if the plots remain within the 5% critical limit, this means that the model parameters are stable and consistent. The model plot shows that CUSUM AT SQUARE is within limits and over time for the Chad data.

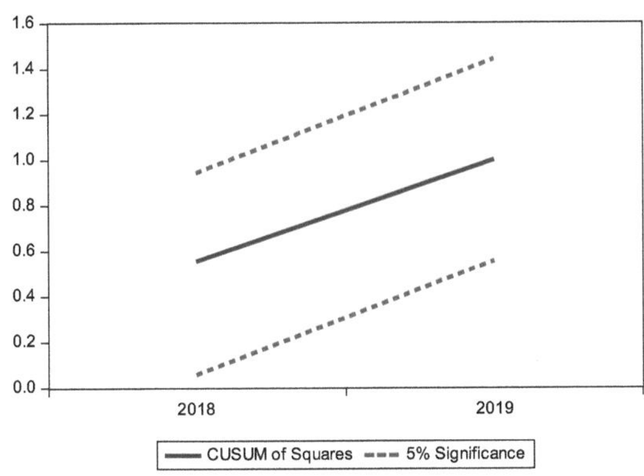

Figure 18 Cusum of squares graph for model 2

iv.   Co-integration test results

Table 20 shows the Co-integration results for model 2. The results of the Co-integration test at the bounds confirm the existence of a Co-integration relationship between the series studied (the value of F-stat is > that of the upper bound); this makes it possible to estimate the long-run effects of diesel consumption, diesel price, vehicle fleet and urbanization on oil rent.

Table 20 Co-integration results for model 2

Statistical test	Value	Significance	I(0)	I(1)
F-statistics	31.33710	10%	1.9	3.01
K	4	5%	2.26	3.48
		2.5%	2.62	3.9
		1%	3.07	4.44

v. Long-term and short-term dynamics

Table 21 shows the long-term results. In fact, the linear impact of diesel consumption on CO_2 emissions is positive and insignificant. In addition, the linear impact of the vehicle fleet, the price of diesel and urbanization remain insignificant.

Table 21 Long-term results for model 2

Variable	Coefficient	Std. Error	t-Statistic	Prob.
LCGA	1.330279	1.235509	1.076705	0.3942
LPGA	-4.607556	9.953909	-0.462889	0.6889
LPAU	-0.509080	0.568267	-0.895847	0.4649
LURB	11.71130	37.89655	0.309034	0.7865

Table 22 shows the short-term results. The error correction term (ECT) is negative and significant; this allows us to say that the error correction model is valid and that emissions of CO_2 emissions are approaching their long-term equilibrium level with an adjustment speed of 19.63%, thanks to the contribution of diesel consumption, the price of diesel, the vehicle fleet and urbanization. In fact, diesel consumption is insignificant. Diesel prices, the car fleet and urbanization, on the other hand, would have a positive and significant linear impact of 1% and 5% respectively. Thus, a 1% increase in the price of diesel, the number of cars on the road and urbanization would lead to an increase in emissions of CO_2 emissions by 2.08%, 0.24% and 5.41% respectively.

Table 22 Short-term results for model 2

Variable	Coefficient	Std. Error	t-Statistic	Prob.
D(LCGA)	0.102556	0.037594	2.727982	0.1122
D(LPGA)	2.083778	0.082962	25.11736	0.0016***
D(LPAU)	0.245617	0.028566	8.598376	0.0133**
D(LURB)	5.415729	0.556560	9.730721	0.0104**
ECT(-1)	-0.196371	0.009057	21.68079	0.0021

Note: ***Significativity at 1%;
**Significativity at 5%.

vi.   Causality between variables in model 2

After estimating the long-term and short-term results, we performed a Granger causality test in the Toda-Yamamoto sense. The results of the Granger causality test in the Toda-Yamamoto sense are shown in Table 23. In this table, three bidirectional causalities emerge: a bidirectional causality between emissions of CO_2 emissions and diesel consumption, a bidirectional causality between CO_2 emissions and the price of diesel, a bidirectional causality between the price of diesel and urbanization; and six unidirectional causalities: a unidirectional causality from the vehicle fleet to emissions of CO_2 emissions, unidirectional causality from CO_2 emissions to urbanization, unidirectional causality from diesel consumption to urbanization, unidirectional causality from diesel consumption to diesel price, unidirectional causality from vehicle fleet to diesel consumption, and unidirectional causality from vehicle fleet to urbanization. Fig. 19 shows the results of Granger causality in the Toda-Yamamoto sense for model 2.

Table 23 Granger causality results in the Toda-Yamamoto sense for model 2

Dependent variables	Explanatory variables					Decisions
	$\ln CO_2$	ln CGA	ln PGA	ln PAU	ln URB	
$\ln CO_2$		6.17***	8.72***	3.21*	0.09	CGA → CO_2 PGA → CO_2 PAU → CO_2
ln CGA	21.90***		1.75	9.03***	1.20	CO_2 → CGA PAU → CGA
ln PGA	45.82***	9.45***		1.89	13.45***	CO_2 → PGA CGA → PGA URB → PGA
ln PAU	2.26	2.48	0.34		0.93	
ln URB	21.45***	3.05*	13.22***	5.15*		CO_2 → URB CGA → URB PGA → URB PAU → URB

Note: (.) : Probability (P-value); ***Significativity at 1%; **Significativity at 5%; *Significativity at 10%.

The → symbol represents unidirectional causality

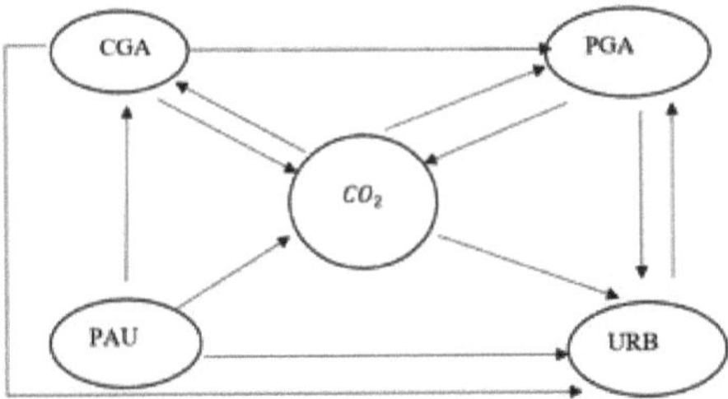

Figure 19 Summary of Granger causality in the Toda-Yamamoto sense

III.2Prediction   of CO_2

Chad wants to be a country that refrains from its greenhouse gas emissions, particularly emissions of CO_2 emissions, with the aim of safeguarding the Paris climate agreements in April 2016 and January 2017. On the basis of this objective, the results for emissions of CO_2 emissions due to the consumption of petroleum products (super and diesel) in the road transport sector for Sequential GMC (1, N) - GA models are presented in the table below. However, the reliability of the results depends on the validation criteria shown in Table 24. Where we note p the number of explanatory variables, $X_1(t)$ represents actual CO_2 emissions for the year, $\widehat{X_1}(t)$ predicted emissions for the year t; N represents the sample size $sd(X_1)$ and $sd(\widehat{X_1})$ represents the standard deviation of X_1 and respectively. $\widehat{X_1}$. Among these different criteria, the measurement criterion par excellence is MAPE.

Table 24 Precision measurements and thresholds

Criteria	Formulas	Threshold levels			
		1er (Mediocre)	2ème (Acceptable)	3ème (Good)	4ème (Perfect)
R^2_{adj}	$1-\dfrac{N-1}{N-P-1}\left(1-\dfrac{\sum_{t-1}^{n}(X_1(t)-\widehat{X_1}(t)^2}{\sum_{t-1}^{n}(X_1(t)-\overline{X_1}(t))^2}\right)$	< 0.90	≥ 0.90	≥ 0.95	≥ 0.98

| R | $\dfrac{1}{N-1}\sum_{t=1}^{N}\dfrac{(X_1(t)-\overline{X_1})(\widehat{X_1}(t)}{\sqrt{sd(X_1)^2 sd(\overline{X}}}$ | < 0.90 | ≥ 0.90 | ≥ 0.95 | ≥ 0.98 |
| $MAPE$ | $\dfrac{1}{N}\sum_{t=1}^{N}\left|\dfrac{X_1(t)-\widehat{X_1}(t)}{X_1(t)}\right|$ | > 0.1 | ≤ 0.1 | ≤ 0.05 | ≤ 0.01 |

Sources : (Hamzacebi and Es, 2014; Sapnken et al., 2022)

Table 25 shows the performance of the Sequential GMC (1, N)-GA models for forecasting emissions of CO_2. It can be seen from Table 25 that the Sequential GMC (1, N) - GA model performs well in predicting emissions of CO_2 emissions in the road transport sector in Chad, with correlation and determination coefficients above 80%. In addition, the Sequential GMC (1, N) - GA model has a MAPE of less than 5%, indicating that the error made when using a predicted value is less than 5%. In addition to the MAPE, Fig. 20 shows that the actual and predicted data are almost identical, which further underlines the model's ability to be more accurate. The Sequential GMC (1, N) - GA model performs better and therefore makes fewer errors when used.

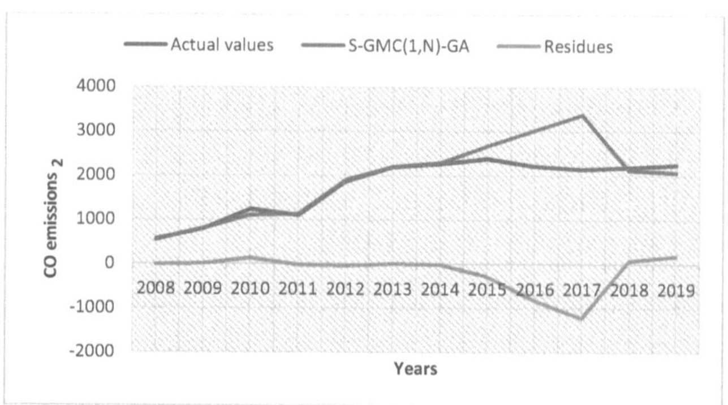

Figure 20:Observed/predicted emissions and residuals for S-GMC (1, N) - GA

Table 25:GMC Sequential Performance (1, N) - GA for forecasting emissions of CO_2

	Data	
Emissions of CO_2 (in kilotonnes)		
Criteria	S-GMC(1, N)-GA	NeuralODE-GM(1,1)
R	0.89	0.85
R^2_{adj}	0.81	0.79
$MAPE$ (%)	1.36	3.28

The results of estimates of CO_2 emissions in the road transport sector, presented in Table 26, generally show that emissions of CO_2 emissions in the road transport sector will increase over time. For the Sequential GMC(1, N)-GA model, emissions of CO_2 emissions in 2030 are estimated at 2672.39 kilotonnes respectively.

Table 26 Forecasted emissions of CO_2

Emissions of CO_2 (in kilotonnes)	S-GMC(1, N)-GA	NeuralODE-GM(1,1)
Years		
2020	2422,03	2417,35
2021	2571,07	2583,21
2022	2120,10	2119,09
2023	2469,14	2475,15
2024	2618,17	2626,72
2025	2527,21	2531,05
2026	2556,25	2549,31
2027	2585,28	2497,17
2028	2614,32	2627,86
2029	2643,35	2646,18
2030	2672,39	2680,20

III.   3Discussion of results

An increase in supermarket consumption would lead to higher emissions of CO_2 emissions in the long term

(Uzair Ali et al., 2022; Umar et al., 2021). Moreover, there is a unidirectional causality from supermarket consumption to emissions of CO_2 emissions (Arora and Kaur, 2020). That said, super consumption could boost emissions of CO_2 emissions in Chad, or alternatively, the more super is consumed, the more emissions of CO_2 emissions increase in the long term. The question of how to reduce CO_2 emissions would therefore become crucial. Chad could therefore seek to reduce its emissions of CO_2 emissions by consuming less super and turning more to alternative energies. This is in line with the work of Nnaji et al (2013). Reducing fossil fuel consumption can be achieved by highlighting new energy efficiency measures while improving existing ones; anything that would save energy in the long term (Umar et al., 2021) or the introduction of policies to reduce fuel consumption by gasoline-powered vehicles would be useful for considerably reducing emissions of greenhouse gases. CO_2 emissions (Solís and Sheinbaum, 2013).. It can also be achieved by controlling the rapid growth of modes of transport without high-quality fuel (Danish et al., 2018). We can refer here to the frelated fuels widely used in Chad when we know that fossil fuels power the vast majority of road vehicles in Chad. It should be noted that the long-term consumption of diesel does not appear to affect emissions of CO_2. This situation can be explained by the price of super diesel, which was generally lower than that of diesel over the study period. The population of Chad would prefer to consume super over diesel, which is more expensive. As a result, the higher consumption of super diesel would lead to an increase in emissions of CO_2. The result of the causality between CO_2 emissions and diesel consumption shows that there is a bidirectional causality between CO_2 emissions and diesel consumption (Mensah et al., 2019; Chandran and Tang, 2013). In fact, the consumption of fossil fuels and emissions of CO_2 emissions are interdependent, i.e. emissions of CO_2 emissions cause diesel consumption (Gokmenoglu and Sadeghieh, 2019) and diesel consumption causes emissions of CO_2 emissions (Khobai and Le Roux, 2017; Kivyiro and Arminen, 2014).. Clearly, higher diesel consumption would lead to higher emissions of CO_2 and vice versa. This result would make it possible to set up tools for controlling diesel consumption in Chad's transport sector, and to use these as a solution for reducing emissions of CO_2 emissions (Mensah et al., 2019; Chandran and Tang, 2013).. However, Chad can use the vehicle fleet and price as tools to control fossil fuel consumption. Moreover, the causality results show us on the one hand that there is a unidirectional causality running from the vehicle fleet to diesel consumption, and on the other a bidirectional relationship between super consumption and the price of super. It should be noted that Chad has considerable renewable energy potential. From north to south, the sun shines from 2,750 to 3,250 hours a year. This gives an average of 4 to 6 kilowatts/hour per square meter per day. (Mbainaissem Peurdoum, n.d.)

The price of super and diesel are significant in the short term at the 1% threshold. In other words, an increase in the price of super and diesel would lead to an increase in emissions in the short term. (A. Malik et al., 2020). In other words, in the short term, the price of super and diesel accelerates emissions of CO_2 . The higher the price of petroleum products in the short term, the greater the increase in CO_2 emissions. Controlling and regulating these prices, combined with a reduction in the use of fossil fuels, could make the negative effects on the environment reversible. (Sadorsky, 2009). This could involve, for example, adjusting domestic petroleum product prices in line with changes in international prices, to bring about a change in consumer behavior (Mahmood et al., 2022). This scenario could reduce emissions by CO_2 emissions in the short term. Moreover, the work of Li et al. (2020) show that prices could play a very important role in reducing greenhouse gas emissions. CO_2. This trend is totally called into question in the long term, where the price of super and diesel remain insignificant. This suggests that, in the long term, emissions of CO_2 emissions are not influenced by any increase in the price of super and diesel, and that Chad should be looking at other causes of CO_2. This result may be based on the causality implications of the figures... which state that consumption of petroleum products (super and diesel) and urbanization influence prices. In other words, if the price of petroleum products (super and diesel) does not influence emissions of CO_2 emissions in the long term, the reason could be the level of consumption of petroleum products and the level of urbanization.

The vehicle fleet is significant in the short term in Models 1 and 2. Thus, a 1% increase in the vehicle fleet would result in a 0.11% increase in emissions of CO_2 emissions for model 1 and a 0.24% increase in CO_2 emissions for model 2. These results show that whatever the petroleum product used (super or diesel), the level of the vehicle fleet would lead to an increase in emissions of CO_2 emissions in the short term. A similar situation is observed in the long term, where in model 1, a 1% increase in the number of cars on the road would raise emissions by CO_2 emissions by 0.36%. In Model 2, on the other hand, the vehicle fleet has no significant impact on emissions of CO_2. These long-term results show that the fleet using premium fuel would increase emissions by CO_2 emissions in the long term, while the diesel-fuelled fleet remains insignificant in the long term, although the causality results show that the diesel-fuelled fleet influences emissions by CO_2. Moreover, in China, the work of Gambhir et al. (2015) show that growth in road transport has led to a considerable increase in emissions of CO_2 emissions in recent decades. These findings are supported by those of Fontaras et al. (2017) who assert that factors such as vehicle configuration and traffic conditions have a major influence on emissions of CO_2 . It is clear that the urgent implementation of a policy to reduce emissions of CO_2 emissions through an adequate vehicle fleet remains imperative.

Finally, urbanization has no significant impact on short- and long-term CO_2 emissions in both the short and long term in Model 1. This result is similar to the work of Danish et al. (2018) in Pakistan. On the other hand, the causality results of the same model show us that emissions of CO_2 emissions influence urbanization. In other words, particular attention should be paid to the speed of emissions of CO_2 emissions, so that urbanization does not necessarily contribute to its increase, as reported in the literature. For model 2, urbanization is also insignificant in terms of emissions of CO_2 emissions in the long term, and is significant in the short term. Thus, in the short term, a 1% increase in urbanization would result in a 5.41% increase in emissions of CO_2. That said, as the population grows, so do emissions of CO_2 emissions. In the long term, however, population growth is insignificant. This result is certainly due to the influence of the vehicle fleet and diesel consumption observed in the causality results shown in table ..... Three quarters of Chad's territory is desert (CDN, 2021). Moreover, with a population of around 16.6 million in 2019, this country has a density of around 12.9 inhabitants per square kilometer, despite a population growth rate of around 3% in 2019 (WATHI, 2021). The country also has a rural population of around 78% and an urban population of almost 21.9%, concentrated mainly in the city of N'Djaména (WATHI, 2021). From the above, we understand that population growth in Chad cannot have a significant impact on emissions of greenhouse gases. CO_2.

Conclusion

In this chapter, we present the results of the estimates of the exogenous variables (super/gasoil consumption, super/gasoil prices, vehicle fleet and urbanization) on the endogenous variable (emissions of CO_2 emissions); to present the results of the causal study of the various variables under study; and finally to highlight future emissions of CO_2 emissions in Chad. To do this, stationarity tests on the different variables showed that they are stationary at different orders I (0) and I (1). The selection criterion for the two models indicated the optimal lag at 1. Subsequently, the estimates of our two models gave us an identical coefficient of determination of 98%, and the robustness tests carried out in both models were verified with p-values above 5% for each of the tests carried out in each model. The Co-integration test of the two models gave us information that both models admit long-term relationships, as F-stat is greater than the upper bounds of each model. As a result, it was deduced that an increase in super and diesel would lead to an increase in emissions of CO_2 . This makes it possible to control the consumption of petroleum products in order to reduce emissions by CO_2 . Secondly, the price of super and diesel are significant in the short term, but in the long term they are insignificant. It is therefore clear that price control and regulation combined with a reduction in the use of fossil fuels could make the negative effects on the environment reversible in the short term, while the level of consumption of petroleum products and urbanization could be factors justifying the insignificant impact in the long term. The car fleet in models 1 and 2 has a significant impact on emissions of CO_2 emissions in the short term. In the long term, it is significant in model 1, but in model 2, it is insignificant, although causality tells us that the car fleet influences emissions of CO_2. For future emissions of CO_2 emissions, the GMC(1, N)-GA sequential model was used, and the results show that by 2030, emissions of CO_2 emissions will increase. The conclusion to be drawn from these results is that the urgent implementation of a policy to reduce emissions of CO_2 emissions requires an adequate vehicle fleet. Finally, urbanization, as reported in the literature (it increases emissions of CO_2 emissions) is not verified in this study. The conclusions to this effect question the influence of the vehicle fleet and diesel consumption in model 2, and recommend a particular look at the speed of emissions of CO_2 emissions in Model 1. The overall conclusion sets out the policy implications for Chadian decision-makers.

GENERAL CONCLUSION AND POLICY IMPLICATIONS

This work examined the linear impact of consumption of petroleum products (super and diesel), prices of petroleum products (super and diesel), the vehicle fleet and urbanization on CO CO_2 emissions; examined the causal links between the various variables studied and explored future emissions of CO_2 emissions in Chad. We therefore proposed four main objectives:

i)   Highlight the short- and long-term dynamics of petroleum product consumption (super and diesel), petroleum product prices (super and diesel), the vehicle fleet and urbanization on emissions of CO_2 emissions in Chad ;

ii)   Provide an analysis of the causal links between emissions of CO_2 emissions, petroleum product consumption, petroleum product prices, the vehicle fleet and urbanization in Chad;

iii)   Predicting future emissions of CO_2 emissions in Chad

iv)   Help political decision-makers in Chad to formulate policies and corresponding measures to ensure the effective control of emissions of CO_2 emissions in the country.

To achieve these objectives, we have studied the emissions of CO_2 emissions according to the sector in which each petroleum product is consumed. In this approach, we have mainly defined the road sector, with super and diesel fuels.

For the road sector, we performed stationarity tests and cointegration tests to confirm the long-term relationships of the different variables. We went on to analyze the short- and long-term dynamics of petroleum product consumption, petroleum product prices, the vehicle fleet and urbanization on CO CO_2 emissions, and the existing causality between CO_2 emissions, petroleum product consumption, prices, vehicle fleet and urbanization for the period 2008 - 2019. To do this, we applied the Augmented Dickey-Fuller and Phillips-Perron tests for the stationarity of variables, the ARDL Bounds approach to test for Co-integration and to identify the short- and long-term effects of exogenous variables that determine emissions of CO_2 emissions in Chad's road sector. Granger causality in the Toda-Yamamoto sense was performed to determine and analyze the various influences between the different variables.

ARDL limit tests confirm the co-integration between the series. Our results show that in the short term, only the price of super and the car fleet have a significant impact on emissions of CO_2 emissions in model 1, and in model 2, diesel prices, the car fleet and urbanization have a significant impact on emissions of CO_2 . In the long term, supermarket consumption and vehicle fleet have a significant impact on emissions of CO_2 emissions, while superoil price and urbanization are insignificant in model 1, and in model 2, no variable has a significant impact on emissions of CO_2. The Granger causality test in the Toda-Yamamoto sense revealed a variety of causalities. In model 1, the causality results indicate that there is bidirectional causality between super consumption and super price, and bidirectional causality between emissions of CO_2 emissions and the price of super. Also, there are five unidirectional causalities, namely unidirectional causality from super consumption to emissions of CO_2 emissions, unidirectional causality from CO_2 emissions to the vehicle fleet, unidirectional causality from CO_2 emissions to urbanization, unidirectional causality from super consumption to vehicle fleet and unidirectional causality from urbanization to vehicle fleet. In model 2, the causality results show that there are three bidirectional causalities, namely a bidirectional causality between emissions of CO_2 emissions and diesel consumption, bidirectional causality between CO_2 emissions and the price of diesel, and a bidirectional causality between the price of diesel and urbanization. There are also six unidirectional causalities, including a unidirectional causality from the vehicle fleet to emissions of CO_2 emissions, unidirectional causality from CO_2 emissions to urbanization, unidirectional causality from diesel consumption to urbanization, unidirectional causality from diesel consumption to diesel price, unidirectional causality from vehicle fleet to diesel consumption and, finally, unidirectional causality from vehicle fleet to urbanization.

With regard to the forecasting of CO_2 emissions in Chad, the GMC (1, N) - GA sequential model was used as the preferred tool for determining the latter. Overall, future emissions of CO_2 emissions in Chad will reach 2672.39 kilotonnes by 2030, reflecting the persistent growth of CO_2 emissions in Chad.

In Chad, emissions of CO_2 emissions are increasing year after year, and are dominated by diesel and super gasoline emissions from the land transport sector. The road transport sector in Chad is growing rapidly, and its vehicles, most of which are obsolete, are the biggest emitters of greenhouse gases. The country has been consuming large quantities of these two petroleum products (diesel and super) for over a decade, and air pollution through emissions of CO_2 emissions. In order to meet the requirements for reducing emissions of CO_2 emissions, the following recommendations are made to the Chadian government:

i) Reducing the consumption of petroleum products, cutting emissions of CO_2 emissions, and steadily increasing the use of clean energy sources such as hydroelectricity, wind power and solar power are just some of the measures that can be taken to reduce carbon emissions.

ii) The best way to significantly reduce CO_2 emissions in cities could be to impose a carbon tax on petroleum-intensive activities. The money raised by this tax would then be used to promote the use of greener energies in these cities.

iii) Putting in place an effective policy would enable industrial and residential consumers to easily install renewable energy sources such as solar power and others within their entities. Such a policy would need to be developed in consultation with leading financial institutions. In any case, it would reduce the amount of fossil fuels purchased by the government.

iv) When oil prices are rising, it would be a good idea to inject the revenues from oil sales into cleaner technologies and production processes. This would help control the environmental consequences of rising oil prices. These environmentally-friendly technologies and production processes also need to be well understood at the time of the agreements between the beneficiary country and the host country, so that the conditions for implementation can be properly defined.

v) Urgently reduce the number of cars on the road that threaten environmental sustainability. The aging fleet, which is in the majority, could be converted to a new, environmentally-friendly fleet.

vi)   Finally, promote a green public transport program
      in cities to reduce dependence on private vehicles
      and engines, which run mainly on diesel and
      premium in Chad. The promotion of fuel-efficient,
      hybrid or gas-powered vehicles could also be
      investigated.

BIBLIOGRAPHICAL REFERENCES

Abdallh, A.A., Abugamos, H., 2017. A semi-parametric panel data analysis onthe urbanisation-carbon emissions nexus for the MENA countries. Renewable and Sustainable Energy Reviews 78, 1350-1356. https://doi.org/10.1016/j.rser.2017.05.006

Adzawla, W., Sawaneh, M., Yusuf, A.M., 2019. Greenhouse gasses emission and economic growth nexus of sub-Saharan Africa. Scientific African 3, e00065.

Ağbulut, Ü., 2022. Forecasting of transportation-related energy demand and CO2 emissions in Turkey with different machine learning algorithms. Sustainable Production and Consumption 29, 141-157. https://doi.org/10.1016/j.spc.2021.10.001

Ahanchian, M., Biona, J.B.M., 2014. Energy demand, emissions forecasts and mitigation strategies modeled over a medium-range horizon: The case of the land transportation sector in Metro Manila. Energy Policy 66, 615-629. https://doi.org/10.1016/j.enpol.2013.11.026

Ahmed, K., Mahalik, M.K., Shahbaz, M., 2016. Dynamics between economic growth, labor, capital and natural resource abundance in Iran: An application of the combined cointegration approach. Resources Policy 49, 213-221. https://doi.org/10.1016/j.resourpol.2016.06.005

IEA, 2021. World Energy Outlook [WWW Document]. URL https://www.iea.org/reports/world-energy-outlook-2021 (accessed 4.6.23).

Akeresola, R.A., Gayawan, E., 2021. Bayesian Spatio-temporal Analysis of Greenhouse Gas Emissions in Africa.

Alam, M.J., Begum, I.A., Buysse, J., Rahman, S., Van Huylenbroeck, G., 2011. Dynamic modeling of causal relationship between energy consumption, CO2 emissions and economic growth in India. Renewable and Sustainable Energy Reviews 15, 3243-3251.

Aldubyan, M., Gasim, A., 2021. Energy price reform in Saudi Arabia: Modeling the economic and environmental impacts and understanding the demand response. Energy Policy 148, 111941. https://doi.org/10.1016/j.enpol.2020.111941

Alkathery, M.A., Chaudhuri, K., 2021. Co-movement between oil price, CO2 emission, renewable energy and energy equities: Evidence from GCC countries. Journal of Environmental Management 297, 113350. https://doi.org/10.1016/j.jenvman.2021.113350

Al-Mulali, U., Ozturk, I., Lean, H.H., 2015. The influence of economic growth, urbanization, trade openness, financial development, and renewable energy on pollution in Europe. Nat Hazards 79, 621-644. https://doi.org/10.1007/s11069-015-1865-9

Ameyaw, B., Yao, L., Oppong, A., Agyeman, J.K., 2019. Investigating, forecasting and proposing emission mitigation pathways for CO2 emissions from fossil fuel combustion only: A case study of selected countries. Energy policy 130, 7-21.

Amiri, A., Ventelou, B., 2012. Granger causality between total expenditure on health and GDP in OECD: Evidence from the Toda-Yamamoto approach. Economics Letters 116, 541-544. https://doi.org/10.1016/j.econlet.2012.04.040

Antonakakis, N., Chatziantoniou, I., Filis, G., 2017. Energy consumption, CO2 emissions, and economic growth: An ethical dilemma. Renewable and Sustainable Energy Reviews 68, 808-824. https://doi.org/10.1016/j.rser.2016.09.105

Ardakani, S.R., Hossein, S.M., Aslani, A., 2018. Statistical approaches to forecasting domestic energy consumption and assessing determinants: The case of Nordic countries. Strategic Planning for Energy and the Environment 38, 26-71.

Arora, R., Kaur, D.B., 2020. Fossil fuel consumption, economic growth and CO2 emissions. Causality evinced from the BRICS world. Theoretical and Applied Economics XXVII, 131-142.

Arouri, M.E.H., Ben Youssef, A., M'henni, H., Rault, C., 2012. Energy consumption, economic growth and CO2 emissions in Middle East and North African countries. Energy Policy 45, 342-349. https://doi.org/10.1016/j.enpol.2012.02.042

ARSAT, 2022. Autorité de Régulation du Secteur Pétrolier Aval du Tchad [WWW Document]. URL http://www.arsat.td/ (accessed 11.2.22).

Azadeh, A., Khakestani, M., Saberi, M., 2009. A flexible fuzzy regression algorithm for forecasting oil consumption estimation. Energy Policy 37, 5567-5579.

Bahrami, S., Hooshmand, R.-A., Parastegari, M., 2014. Short term electric load forecasting by wavelet transform and grey model improved by PSO (particle swarm optimization) algorithm. Energy 72, 434-442. https://doi.org/10.1016/j.energy.2014.05.065

World Bank, 2022. Urban population growth (annual %) - Chad | Data [WWW Document]. URL https://donnees.banquemondiale.org/indicateur/SP. URB.GROW?locations=TD (accessed 1.19.23).

Bedoum, A., Bouka Biona, C., Jean Pierre, B., Adoum, I., Mbiake, R., Baohoutou, L., 2017. Evolution of climate extremes indices in the Republic of Chad from 1960 to 2008. Atmosphere-Ocean 55, 42-56. https://doi.org/10.1080/07055900.2016.1268995

Belbute, J.M., Pereira, A.M., 2020. Reference forecasts for CO2 emissions from fossil-fuel combustion and cement production in Portugal. Energy policy 144, 111642.

Bewket, W., 2012. Climate change perceptions and adaptive responses of smallholder farmers in central highlands of Ethiopia. International Journal of Environmental Studies 69, 507-523. https://doi.org/10.1080/00207233.2012.683328

Böhringer, C., Rutherford, T.F., Schneider, J., 2021. The incidence of CO2 emissions pricing under alternative international market responses: A computable general equilibrium analysis for Germany. Energy Economics 101, 105404. https://doi.org/10.1016/j.eneco.2021.105404

Brown, R.L., Durbin, J., Evans, J.M., 1975. Techniques for Testing the Constancy of Regression Relationships Over Time. Journal of the Royal Statistical Society : Series B (Methodological) 37, 149-163. https://doi.org/10.1111/j.2517-6161.1975.tb01532.x

Carbonnier, G., Grinevald, J., 2011. Energy and Development, in: Carbonnier, G. (Ed.), International Development Policy: Energy and Development. Palgrave Macmillan UK, London, pp. 3-20. https://doi.org/10.1007/978-0-230-31401-6_1

UNFCCC, 2001. High National Committee for the Environment.

CDN, 2021. Update of the national determined contribution, Republic of Chad.

Chandran, V.G.R., Tang, C.F., 2013. The impacts of transport energy consumption, foreign direct investment and income on CO2 emissions in ASEAN-5 economies. Renewable and Sustainable Energy Reviews 24, 445-453. https://doi.org/10.1016/j.rser.2013.03.054

Change, G.C., 2008. Livestock and Global Climate Change.Cambridge university press.

CHANGE, I.P.O.C., 2007. REPORT OF THE NINETEENTH SESSION OF THE INTERGOVERNMENTAL PANEL ON CLIMATE CHANGE (IPCC) Geneva, 17-20 (am only) April 2002.

Chen, C.-I., Huang, S.-J., 2013. The necessary and sufficient condition for GM (1, 1) grey prediction model. Applied Mathematics and Computation 219, 6152-6162.

Chen, S.-S., Xu, J.-H., Fan, Y., 2015. Evaluating the effect of coal mine safety supervision system policy in China's coal mining industry: A two-phase analysis. Resources Policy 46, 12-21.

Cherniwchan, J., 2012. Economic growth, industrialization, and the environment. Resource and Energy Economics 34, 442-467. https://doi.org/10.1016/j.reseneeco.2012.04.004

Cotte Poveda, A., Pardo Martínez, C.I., 2011. Trends in economic growth, poverty and energy in Colombia: long-run and short-run effects. Energy Systems 2, 281-298.

Danish, Baloch, M.A., Suad, S., 2018. Modeling the impact of transport energy consumption on CO2 emission in Pakistan: Evidence from ARDL approach. Environ Sci Pollut Res 25, 9461-9473. https://doi.org/10.1007/s11356-018-1230-0

De Cáceres, A.M., Martínez, P.J.P., 2009. Transporte y cambio climático en España: problemas y perspectivas. Carreteras 4, 44-50.

Deng, J.L., 1982. Control problems of grey systems. Systems & control letters 1, 288-294.

Deng, J.-L., 1982. Control problems of grey systems. Systems & control letters 1, 288-294.

Dengiz, A.Ö., Atalay, K.D., Dengiz, O., 2018. Grey forecasting model for CO 2 emissions of developed countries, in: The International Symposium for Production Research. Springer, pp. 604-611.

Dickey, D.A., Fuller, W.A., 1979. Distribution of the Estimators for Autoregressive Time Series with a Unit Root. Journal of the American Statistical Association 74, 427-431. https://doi.org/10.1080/01621459.1979.10482531

Ding, Song, Dang, Y.-G., Li, X.-M., Wang, J.-J., Zhao, K., 2017. Forecasting Chinese CO2 emissions from fuel combustion using a novel grey multivariable model. Journal of Cleaner Production 162, 1527-1538.

Ding, S., Dang, Y.G., Xu, N., Wei, L., Ye, J., 2017. Multi-variable time-delayed discrete grey model. Control and decision 32, 1997-2004.

Ding, S., Hipel, K.W., Dang, Y., 2018. Forecasting China's electricity consumption using a new grey prediction model. Energy 149, 314-328. https://doi.org/10.1016/j.energy.2018.01.169

Ding, S., Zhang, M., Song, Y., 2019. Exploring China's carbon emissions peak for different carbon tax scenarios. Energy Policy 129, 1245-1252. https://doi.org/10.1016/j.enpol.2019.03.037

Dong, F., Wang, Y., Su, B., Hua, Y., Zhang, Y., 2019. The process of peak CO_2 emissions in developed economies: A perspective of industrialization and urbanization. Resources, Conservation and Recycling 141, 61-75. https://doi.org/10.1016/j.resconrec.2018.10.010

Dong, H., Xue, M., Xiao, Y., Liu, Y., 2021. Do carbon emissions impact the health of residents? Considering China's industrialization and urbanization. Science of The Total Environment 758, 143688. https://doi.org/10.1016/j.scitotenv.2020.143688

Du, H., Liu, D., Southworth, F., Ma, S., Qiu, F., 2017. Pathways for energy conservation and emissions mitigation in road transport up to 2030: A case study of the Jing-Jin-Ji area, China. Journal of Cleaner Production 162, 882-893. https://doi.org/10.1016/j.jclepro.2017.06.054

Duan, L., Hu, W., Deng, D., Fang, W., Xiong, M., Lu, P., Li, Z., Zhai, C., 2021. Impacts of reducing air pollutants and CO_2 emissions in urban road transport through 2035 in Chongqing, China. Environmental Science and Ecotechnology 8, 100125. https://doi.org/10.1016/j.ese.2021.100125

Ehrenberger, S., Seum, S., Pregger, T., Simon, S., Knitschky, G., Kugler, U., 2021. Land transport development in three integrated scenarios for

Germany - Technology options, energy demand and emissions. Transportation Research Part D: Transport and Environment 90, 102669. https://doi.org/10.1016/j.trd.2020.102669

Engle, R.F., Granger, C.W.J., 1987. Co-Integration and Error Correction: Representation, Estimation, and Testing. Econometrica 55, 251-276. https://doi.org/10.2307/1913236

Engo, J., 2019. Decoupling analysis of CO2 emissions from transport sector in Cameroon. Sustainable Cities and Society 51, 101732.

Fatima, N., Li, Y., Ahmad, M., Jabeen, G., Li, X., 2021. Factors influencing renewable energy generation development: a way to environmental sustainability. Environ Sci Pollut Res 28, 51714-51732. https://doi.org/10.1007/s11356-021-14256-z

Fontaras, G., Zacharof, N.-G., Ciuffo, B., 2017. Fuel consumption and CO2 emissions from passenger cars in Europe - Laboratory versus real-world emissions. Progress in Energy and Combustion Science 60, 97-131. https://doi.org/10.1016/j.pecs.2016.12.004

Gambhir, A., Tse, L.K.C., Tong, D., Martinez-Botas, R., 2015. Reducing China's road transport sector CO2 emissions to 2050: Technologies, costs and decomposition analysis. Applied Energy 157, 905-917. https://doi.org/10.1016/j.apenergy.2015.01.018

Gokmenoglu, K.K., Sadeghieh, M., 2019. Financial Development, CO2 Emissions, Fossil Fuel Consumption and Economic Growth: The Case of Turkey. Strategic Planning for Energy and the Environment 38, 7-28. https://doi.org/10.1080/10485236.2019.12054409

González Palencia, J.C., Furubayashi, T., Nakata, T., 2012. Energy use and CO2 emissions reduction potential in passenger car fleet using zero emission vehicles and lightweight materials. Energy, 6th Dubrovnik Conference on Sustainable Development of Energy Water and Environmental Systems, SDEWES 2011 48, 548-565. https://doi.org/10.1016/j.energy.2012.09.041

Grodzicki, T., Jankiewicz, M., 2022. The impact of renewable energy and urbanization on CO2 emissions in Europe - Spatio-temporal approach. Environmental Development 44, 100755. https://doi.org/10.1016/j.envdev.2022.100755

Guilyardi, E., Lescarmontier, L., Matthews, R., Point, S.P., Rumjaun, A.B., Schlüpmann, J., Wilgenbus, D., 2018. IPCC Special Report "Global Warming of 1.5° C": Summary for Teachers.

Guo, H., Deng, S., Yang, J., Liu, J., Nie, C., 2020. Analysis and prediction of industrial energy conservation in underdeveloped regions of China using a data pre-processing grey model. Energy Policy 139, 111244.

Guo, J., Zhang, Y.-J., Zhang, K.-B., 2018. The key sectors for energy conservation and carbon emissions reduction in China: evidence from the input-output method. Journal of Cleaner Production 179, 180-190.

Hammoudeh, S., Nguyen, D.K., Sousa, R.M., 2014. Energy prices and CO2 emission allowance prices: A quantile regression approach. Energy Policy 70, 201-206. https://doi.org/10.1016/j.enpol.2014.03.026

Hamzacebi, C., Es, H.A., 2014a. Forecasting the annual electricity consumption of Turkey using an optimized grey model. Energy 70, 165-171.

Hamzacebi, C., Es, H.A., 2014b. Forecasting the annual electricity consumption of Turkey using an optimized grey model. Energy 70, 165-171. https://doi.org/10.1016/j.energy.2014.03.105

Hamzacebi, C., Es, H.A., 2014c. Forecasting the annual electricity consumption of Turkey using an optimized grey model. Energy 70, 165-171.

Hamzacebi, C., Es, H.A., 2014d. Forecasting the annual electricity consumption of Turkey using an optimized grey model. Energy 70, 165-171.

Hamzacebi, C., Karakurt, I., 2015. Forecasting the energy-related CO2 emissions of Turkey using a grey prediction model. Energy Sources, Part A: Recovery, Utilization, and Environmental Effects 37, 1023-1031.

Hansen, C., Daim, T., Ernst, H., Herstatt, C., 2016. The future of rail automation: A scenario-based technology roadmap for the rail automation market. Technological Forecasting and Social Change 110, 196-212. https://doi.org/10.1016/j.techfore.2015.12.017

Hong, T., Jeong, K., Koo, C., 2018. An optimized gene expression programming model for forecasting the national CO2 emissions in 2030 using the metaheuristic algorithms. Applied energy 228, 808-820.

Hossain, M.S., 2011a. Panel estimation for CO2 emissions, energy consumption, economic growth, trade openness and urbanization of newly industrialized countries. Energy policy 39, 6991-6999.

Hossain, M.S., 2011b. Panel estimation for CO2 emissions, energy consumption, economic growth, trade openness and urbanization of newly industrialized countries. Energy policy 39, 6991-6999.

Hosseini, S.M., Saifoddin, A., Shirmohammadi, R., Aslani, A., 2019. Forecasting of CO2 emissions in Iran based on time series and regression analysis. Energy Reports 5, 619-631.

Hu, Y.-C., Jiang, P., Tsai, J.-F., Yu, C.-Y., 2021. An optimized fractional grey prediction model for carbon dioxide emissions forecasting. International Journal of Environmental Research and Public Health 18, 587.

Indonesia, P.R., 1999. Peraturan Pemerintah No. 41 Tahun 1999 Tentang: Pengendalian Pencemaran Udara. No. 41, 1-34.

Jabeen, G., Ahmad, M., Zhang, Q., 2021. Perceived critical factors affecting consumers' intention to purchase renewable generation technologies: rural-urban heterogeneity. Energy 218, 119494. https://doi.org/10.1016/j.energy.2020.119494

Jang, J.-S., 1993. ANFIS: adaptive-network-based fuzzy inference system. IEEE transactions on systems, man, and cybernetics 23, 665-685.

Ji, Q., Zhang, D., Geng, J., 2018. Information linkage, dynamic spillovers in prices and volatility between the carbon and energy markets. Journal of Cleaner Production 198, 972-978. https://doi.org/10.1016/j.jclepro.2018.07.126

Johansen, S., 1991. Estimation and Hypothesis Testing of Cointegration Vectors in Gaussian Vector Autoregressive Models. Econometrica 59, 1551-1580. https://doi.org/10.2307/2938278

Johansen, S., 1988. Statistical analysis of cointegration vectors. Journal of Economic Dynamics and Control 12, 231-254. https://doi.org/10.1016/0165-1889(88)90041-3

Josephine, K.W.N., 2007. Impact of climate change on agriculture in Africa by 2030. Scientific Research and Essays 2, 238-243.

Kayacan, E., Ulutas, B., Kaynak, O., 2010. Grey system theory-based models in time series prediction. Expert systems with applications 37, 1784-1789.

Kaynar, O., Yilmaz, I., Demirkoparan, F., 2011. Forecasting of natural gas consumption with neural network and neuro fuzzy system. Energy Education Science and Technology Part A: Energy Science and Research 26, 221-238.

Khobai, H.B., Le Roux, P., 2017. The relationship between energy consumption, economic growth and carbon dioxide emission: The case of South Africa. International Journal of Energy Economics and Policy 7, 102-109.

Kivyiro, P., Arminen, H., 2014. Carbon dioxide emissions, energy consumption, economic growth, and foreign direct investment: Causality analysis for Sub-Saharan Africa. Energy 74, 595-606.

Köne, A.Ç., Büke, T., 2010. Forecasting of CO2 emissions from fuel combustion using trend analysis. Renewable and Sustainable Energy Reviews 14, 2906-2915.

Kubáňová, J., Kubasáková, I., Dočkalik, M., 2021. Analysis of the Vehicle Fleet in the EU with Regard to Emissions Standards. Transportation Research Procedia, International Scientific Conference "Horizons of Railway Transport 2020" 53, 180-187. https://doi.org/10.1016/j.trpro.2021.02.024

Leimbach, M., Roming, N., Schultes, A., Schwerhoff, G., 2018. Long-Term Development Perspectives of Sub-Saharan Africa under Climate Policies. Ecological Economics 144, 148-159. https://doi.org/10.1016/j.ecolecon.2017.07.033

Li, A., Peng, D., Wang, D., Yao, X., 2017. Comparing regional effects of climate policies to promote non-fossil fuels in China. Energy 141, 1998-2012.

Li, B., Haneklaus, N., 2022. Reducing CO2 emissions in G7 countries: The role of clean energy consumption, trade openness and urbanization. Energy Reports, 2021 International Conference on New Energy and Power Engineering 8, 704-713. https://doi.org/10.1016/j.egyr.2022.01.238

Li, K., Fang, L., He, L., 2020. The impact of energy price on CO2 emissions in China: A spatial econometric analysis. Science of The Total Environment 706, 135942. https://doi.org/10.1016/j.scitotenv.2019.135942

Liu, Z., Deng, Z., Davis, S.J., Giron, C., Ciais, P., 2022. Monitoring global carbon emissions in 2021. Nat Rev Earth Environ 3, 217-219. https://doi.org/10.1038/s43017-022-00285-w

Lotfalipour, M.R., Falahi, M.A., Ashena, M., 2010. Economic growth, CO2 emissions, and fossil fuels consumption in Iran. Energy 35, 5115-5120. https://doi.org/10.1016/j.energy.2010.08.004

Mahmood, H., Alkhateeb, T.T.Y., Furqan, M., 2020. Oil sector and CO2 emissions in Saudi Arabia: asymmetry analysis. Palgrave Commun 6, 1-10. https://doi.org/10.1057/s41599-020-0470-z

Mahmood, H., Asadov, A., Tanveer, M., Furqan, M., Yu, Z., 2022. Impact of Oil Price, Economic Growth and Urbanization on CO2 Emissions in GCC Countries: Asymmetry Analysis. Sustainability 14, 4562. https://doi.org/10.3390/su14084562

Majeed, A., Wang, L., Zhang, X., Muniba, Kirikkaleli, D., 2021. Modeling the dynamic links among natural resources, economic globalization, disaggregated energy consumption, and environmental quality: Fresh evidence from GCC economies. Resources Policy 73, 102204. https://doi.org/10.1016/j.resourpol.2021.102204

Malik, A., Hussain, E., Baig, S., Khokhar, M.F., 2020. Forecasting CO2 emissions from energy consumption in Pakistan under different scenarios: The China-Pakistan economic corridor. Greenhouse Gases: Science and Technology 10, 380-389.

Malik, M.Y., Latif, K., Khan, Z., Butt, H.D., Hussain, M., Nadeem, M.A., 2020. Symmetric and asymmetric impact of oil price, FDI and economic growth on carbon emission in Pakistan: Evidence from ARDL and non-linear ARDL approach. Science of The Total Environment 726, 138421. https://doi.org/10.1016/j.scitotenv.2020.138421

Mardani, A., Streimikiene, D., Nilashi, M., Arias Aranda, D., Loganathan, N., Jusoh, A., 2018. Energy consumption, economic growth, and CO2 emissions in G20 countries: application of adaptive neuro-fuzzy inference system. Energies 11, 2771.

Martínez-Jaramillo, J.E., Arango-Aramburo, S., Álvarez-Uribe, K.C., Jaramillo-Álvarez, P., 2017. Assessing the impacts of transport policies through energy system simulation: The case of the Medellin

Metropolitan Area, Colombia. Energy Policy 101, 101-108. https://doi.org/10.1016/j.enpol.2016.11.026

Mavrotas, G., Kelly, R., 2001. Old wine in new bottles: Testing causality between savings and growth. The Manchester School 69, 97-105.

Mbainaissem Peurdoum, R., n.d. RAPPORT NATIONAL DU TCHAD [WWW Document]. URL https://docplayer.fr/13917919-Rapport-national-du-tchad.html (accessed 3.26.23).

Meng, G., Guo, Z., Li, J., 2021. The dynamic linkage among urbanization, industrialization and carbon emissions in China: Insights from spatiotemporal effect. Science of The Total Environment 760, 144042. https://doi.org/10.1016/j.scitotenv.2020.144042

Meng, M., Niu, D., 2011. Modeling CO_2 emissions from fossil fuel combustion using the logistic equation. Energy 36, 3355-3359.

Meng, M., Niu, D., Shang, W., 2012. CO_2 emissions and economic development: China's 12th five-year plan. Energy Policy 42, 468-475.

Mensah, I.A., Sun, M., Gao, C., Omari-Sasu, A.Y., Zhu, D., Ampimah, B.C., Quarcoo, A., 2019. Analysis on the nexus of economic growth, fossil fuel energy consumption, CO_2 emissions and oil price in Africa based on a PMG panel ARDL approach. Journal of Cleaner Production 228, 161-174. https://doi.org/10.1016/j.jclepro.2019.04.281

MESTECC, 2018. Malaysia: Third National Communication and Second Biennial Update Report to the UNFCCC.

Miller, G.T., Spoolman, S., 2011. Living in the environment: principles, connections, and solutions. Cengage Learning.

Narayan, P.K., 2005. The saving and investment nexus for China: evidence from cointegration tests. Applied Economics 37, 1979-1990. https://doi.org/10.1080/00036840500278103

Nnaji, C.E., Chukwu, J.O., Nnaji, M., 2013. Electricity Supply, Fossil fuel Consumption, Co2 Emissions and Economic Growth: Implications and Policy Options for Sustainable Development in Nigeria. International Journal of Energy Economics and Policy 3, 262-271.

Odhiambo, N.M., 2009. Energy consumption and economic growth nexus in Tanzania: An ARDL bounds testing approach. Energy Policy 37, 617-622. https://doi.org/10.1016/j.enpol.2008.09.077

Ofosu-Adarkwa, J., Xie, N., Javed, S.A., 2020. Forecasting CO2 emissions of China's cement industry using a hybrid Verhulst-GM (1, N) model and emissions' technical conversion. Renewable and Sustainable Energy Reviews 130, 109945.

Onat, N.C., Kucukvar, M., Tatari, O., 2014. Towards life cycle sustainability assessment of alternative passenger vehicles. Sustainability 6, 9305-9342.

World Meteorological Organization, 2021. Climate change in 2020: increasingly alarming indicators and impacts [WWW Document]. URL https://public.wmo.int/fr/medias/communiqu%C3%A9s-de-presse/changement-climatique-en-2020-des-indicateurs-et-des-effets-de-plus-en (accessed 9.19.22).

Ozcan, B., 2013. The nexus between carbon emissions, energy consumption and economic growth in Middle East countries: a panel data analysis. Energy Policy 62, 1138-1147.

Özmen, A., Yılmaz, Y., Weber, G.-W., 2018. Natural gas consumption forecast with MARS and CMARS models for residential users. Energy Economics 70, 357-381.

Pachauri, R.K., Meyer, L.A., 2014. Climate Change 2014: Synthesis Report. Contribution of Working Groups I, II and III to the Fifth Assessment Report of the Intergovernmental Panel on Climate Change.

Pai, T.-Y., Lo, H.-M., Wan, T.-J., Chen, L., Hung, P.-S., Lo, H.-H., Lai, W.-J., Lee, H.-Y., 2015. Predicting air pollutant emissions from a medical incinerator using grey model and neural network. Applied Mathematical Modelling 39, 1513-1525.

Pao, H.-T., Fu, H.-C., Tseng, C.-L., 2012. Forecasting of CO2 emissions, energy consumption and economic growth in China using an improved grey model. Energy 40, 400-409.

Pérez-Suárez, R., López-Menéndez, A.J., 2015. Growing green? Forecasting CO2 emissions with environmental Kuznets curves and logistic growth models. Environmental Science & Policy 54, 428-437.

Pesaran, M.H., Shin, Y., Smith, R.J., 2001. Bounds testing approaches to the analysis of level relationships. J. Appl. Econ. 16, 289-326. https://doi.org/10.1002/jae.616

Piecyk, M.I., McKinnon, A.C., 2010. Forecasting the carbon footprint of road freight transport in 2020. International Journal of Production Economics 128, 31-42.

Pita, P., Chunark, P., Limmeechokchai, B., 2017. CO2 Reduction Perspective in Thailand's Transport sector towards 2030. Energy Procedia 138, 635-640.

Pita, P., Winyuchakrit, P., Limmeechokchai, B., 2020. Analysis of factors affecting energy consumption and CO2 emissions in Thailand's road passenger transport. Heliyon 6, e05112. https://doi.org/10.1016/j.heliyon.2020.e05112

Poumanyvong, P., Kaneko, S., 2010. Does urbanization lead to less energy use and lower CO2 emissions? A cross-country analysis. Ecological Economics, Special Section: Ecological Distribution Conflicts 70, 434-444. https://doi.org/10.1016/j.ecolecon.2010.09.029

Qiao, W., Lu, H., Zhou, G., Azimi, M., Yang, Q., Tian, W., 2020. A hybrid algorithm for carbon dioxide emissions forecasting based on improved lion swarm optimizer. Journal of Cleaner Production 244, 118612.

Qiao, X., Xiao, W., Jaffe, D., Kota, S.H., Ying, Q., Tang, Y., 2015. Atmospheric wet deposition of sulfur and nitrogen in Jiuzhaigou national nature reserve, Sichuan province, China. Science of the Total Environment 511, 28-36.

Sadorsky, P., 2009. Renewable energy consumption, CO2 emissions and oil prices in the G7 countries. Energy Economics 31, 456-462. https://doi.org/10.1016/j.eneco.2008.12.010

Santos, G., 2017a. Road transport and CO2 emissions: What are the challenges? Transport Policy 59, 71-74. https://doi.org/10.1016/j.tranpol.2017.06.007

Santos, G., 2017b. Road transport and CO2 emissions: What are the challenges? Transport Policy 59, 71-74. https://doi.org/10.1016/j.tranpol.2017.06.007

Sapnken, E.F., 2018. Modeling and forecasting gasoline consumption in Cameroon using linear regression models, in: Modeling and Forecasting Gasoline Consumption in Cameroon Using Linear Regression Models: Sapnken, Emmanuel Flavian.

Sapnken, F.E., 2023. A new hybrid multivariate grey model based on genetic algorithms optimization and its application in forecasting oil products demand. Grey Systems: Theory and Application 13, 406-420.

Sapnken, F.E., Ahmat, K.A., Boukar, M., Biobiongono Nyobe, S.L., Tamba, J.G., 2022. Forecasting petroleum products consumption in Cameroon's household sector using a sequential GMC (1, n) model optimized by genetic algorithms. Heliyon 8, e12138. https://doi.org/10.1016/j.heliyon.2022.e12138

Sapnken, F.E., Tamba, J.G., 2022. Petroleum products consumption forecasting based on a new structural auto-adaptive intelligent grey prediction model. Expert Systems with Applications 203, 117579.

Sawitri, E., Hardiman, G., Buchori, I., 2017. The difference of level CO2 emissions from the transportation sector between weekdays and weekend days on the City

Centre of Pemalang. IOP Conf. Ser: Earth Environ. Sci. 70, 012010. https://doi.org/10.1088/1755-1315/70/1/012010

Schipper, L., Saenger, C., Sudardshan, A., 2011. Transport and Carbon Emissions in the United States: The Long View. Energies 4, 563-581. https://doi.org/10.3390/en4040563

Shahbaz et al. 2014. Economic growth, electricity consumption, urbanization and environmental degradation relationship in United Arab Emirates. Ecological Indicators 45, 622-631. https://doi.org/10.1016/j.ecolind.2014.05.022

Shahbaz, M., Loganathan, N., Muzaffar, A.T., Ahmed, K., Ali Jabran, M., 2016. How urbanization affects CO_2 emissions in Malaysia? The application of STIRPAT model. Renewable and Sustainable Energy Reviews 57, 83-93. https://doi.org/10.1016/j.rser.2015.12.096

Shan, S., Ahmad, M., Tan, Z., Adebayo, T.S., Man Li, R.Y., Kirikkaleli, D., 2021. The role of energy prices and non-linear fiscal decentralization in limiting carbon emissions: Tracking environmental sustainability. Energy 234, 121243. https://doi.org/10.1016/j.energy.2021.121243

Shen, Q.-Q., Cao, Y., Yao, L.-Q., Zhu, Z.-K., 2021. An optimized discrete grey multi-variable convolution model and its applications. Comp. Appl. Math. 40, 58. https://doi.org/10.1007/s40314-021-01448-z

Sobrino, N., Monzon, A., 2014. The impact of the economic crisis and policy actions on GHG emissions from road transport in Spain. Energy Policy 74, 486-498. https://doi.org/10.1016/j.enpol.2014.07.020

Solaymani, S., 2019. CO_2 emissions patterns in 7 top carbon emitter economies: The case of transport sector. Energy 168, 989-1001. https://doi.org/10.1016/j.energy.2018.11.145

Solaymani, S., Kardooni, R., Kari, F., Yusoff, S.B., 2015. Economic and environmental impacts of energy subsidy reform and oil price shock on the Malaysian transport sector. Travel Behaviour and Society 2, 65-77.

Solís, J.C., Sheinbaum, C., 2013. Energy consumption and greenhouse gas emission trends in Mexican road transport. Energy for Sustainable Development 17, 280-287. https://doi.org/10.1016/j.esd.2012.12.001

Sun, W., Liu, M., 2016. Prediction and analysis of the three major industries and residential consumption CO_2 emissions based on least squares support vector machine in China. Journal of Cleaner Production 122, 144-153.

Sun, W., Wang, C., Zhang, C., 2017. Factor analysis and forecasting of CO_2 emissions in Hebei, using extreme learning machine based on particle swarm optimization. Journal of cleaner production 162, 1095-1101.

Tamba, J.G., Njomo, D., Limanond, T., Ntsafack, B., 2012. Causality analysis of diesel consumption and economic growth in Cameroon. Energy Policy 45, 567-575. https://doi.org/10.1016/j.enpol.2012.03.006

TCNTCC, 2020. Third Chad National Communication on Climate Change, Republic of Chad.

Teter, J., 2020. Tracking transport 2020.

Tien, T.-L., 2012. A research on the grey prediction model GM (1, n). Applied mathematics and computation 218, 4903-4916.

Toda, H.Y., Yamamoto, T., 1995. Statistical inference in vector autoregressions with possibly integrated processes. Journal of Econometrics 66, 225-250. https://doi.org/10.1016/0304-4076(94)01616-8

Tseng, F.-M., Yu, H.-C., Tzeng, G.-H., 2001. Applied Hybrid Grey Model to Forecast Seasonal Time Series. Technological Forecasting and Social Change 67, 291-302. https://doi.org/10.1016/S0040-1625(99)00098-0

Tsiakmakis, S., Fontaras, G., Ciuffo, B., Samaras, Z., 2017. A simulation-based methodology for quantifying European passenger car fleet CO_2

emissions. Applied Energy 199, 447-465. https://doi.org/10.1016/j.apenergy.2017.04.045

Udemba, E.N., Yalçıntaş, S., 2021. Interacting force of foreign direct invest (FDI), natural resource and economic growth in determining environmental performance: A nonlinear autoregressive distributed lag (NARDL) approach. Resources Policy 73, 102168. https://doi.org/10.1016/j.resourpol.2021.102168

Ulucak, R., Khan, S.U.-D., 2020. Determinants of the ecological footprint: role of renewable energy, natural resources, and urbanization. Sustainable Cities and Society 54, 101996.

Umar, M., Ji, X., Kirikkaleli, D., Alola, A.A., 2021. The imperativeness of environmental quality in the United States transportation sector amidst biomass-fossil energy consumption and growth. Journal of Cleaner Production 285, 124863. https://doi.org/10.1016/j.jclepro.2020.124863

UN, 1992. The United Nations Framework Convention on Climate Change 1 Review of European, Comparative & International Environmental Law 1992 [WWW Document]. URL https://heinonline.org/HOL/LandingPage?handle=hei n.journals/reel1&div=50&id=&page= (accessed 9.28.22).

UNFCCC, U.F.C. on C.C., 2009. Kyoto Protocol Reference Manual on Accounting of Emissions and Assigned Amount (Working Paper). eSocialSciences.

Uzair Ali, M., Gong, Z., Ali, M.U., Asmi, F., Muhammad, R., 2022. CO2 emission, economic development, fossil fuel consumption and population density in India, Pakistan and Bangladesh: A panel investigation. International Journal of Finance & Economics 27, 18-31. https://doi.org/10.1002/ijfe.2134

Van der Zwaan, B., Kober, T., Dalla Longa, F., van der Laan, A., Kramer, G.J., 2018. An integrated assessment of pathways for low-carbon development in Africa. Energy Policy 117, 387-395.

Wang, K.-H., Su, C.-W., Lobonţ, O.-R., Umar, M., 2021. Whether crude oil dependence and CO_2 emissions influence military expenditure in net oil importing countries? Energy Policy 153, 112281. https://doi.org/10.1016/j.enpol.2021.112281

Wang, Q., Li, S., Li, R., 2018. Forecasting energy demand in China and India: Using single-linear, hybrid-linear, and non-linear time series forecast techniques. Energy 161, 821-831. https://doi.org/10.1016/j.energy.2018.07.168

Wang, Q., Li, S., Pisarenko, Z., 2020. Modeling carbon emission trajectory of China, US and India. Journal of Cleaner Production 258, 120723. https://doi.org/10.1016/j.jclepro.2020.120723

Wang, S., Li, G., Fang, C., 2018a. Urbanization, economic growth, energy consumption, and CO_2 emissions: Empirical evidence from countries with different income levels. Renewable and Sustainable Energy Reviews 81, 2144-2159. https://doi.org/10.1016/j.rser.2017.06.025

Wang, S., Zeng, J., Huang, Y., Shi, C., Zhan, P., 2018b. The effects of urbanization on CO_2 emissions in the Pearl River Delta: A comprehensive assessment and panel data analysis. Applied Energy 228, 1693-1706. https://doi.org/10.1016/j.apenergy.2018.06.155

Wang, Z.-X., Ye, D.-J., 2017. Forecasting Chinese carbon emissions from fossil energy consumption using non-linear grey multivariable models. Journal of Cleaner Production 142, 600-612.

WATHI, 2021. General presentation of Chad. WATHI. URL https://www.wathi.org/contexte-election-tchad-2021/presentation-generale-du-tchad/ (accessed 3.26.23).

Wen, L., Cao, Y., 2020. Influencing factors analysis and forecasting of residential energy-related CO_2

emissions utilizing optimized support vector machine. Journal of Cleaner Production 250, 119492.

Wolde-Rufael, Y., 2005. Energy demand and economic growth: the African experience. Journal of Policy Modeling 27, 891-903.

Wolde-Rufael, Y., 2004. Disaggregated industrial energy consumption and GDP: the case of Shanghai, 1952-1999. Energy economics 26, 69-75.

World Resources Institute, 2022. Data Explorer | Climate Watch [WWW Document]. URL https://www.climatewatchdata.org/data-explorer/historical-emissions?historical-emissions-data-sources=cait&historical-emissions-gases=all-ghg&historical-emissions-regions=All%20Selected&historical-emissions-sectors=total-excluding-lucf&page=1 (accessed 9.19.22).

Wu, L., Liu, S., Fang, Z., Xu, H., 2015a. Properties of the GM (1, 1) with fractional order accumulation. Applied Mathematics and Computation 252, 287-293.

Wu, L., Liu, S., Liu, D., Fang, Z., Xu, H., 2015b. Modelling and forecasting CO2 emissions in the BRICS (Brazil, Russia, India, China, and South Africa) countries using a novel multi-variable grey model. Energy 79, 489-495.

Xie, M., Yan, S., Wu, L., Liu, Liying, Bai, Y., Liu, Linghui, Tong, Y., 2021. A novel robust reweighted multivariate grey model for forecasting the greenhouse gas emissions. Journal of Cleaner Production 292, 126001.

Xie, W., Wu, W.-Z., Liu, C., Zhang, T., Dong, Z., 2021. Forecasting fuel combustion-related CO2 emissions by a novel continuous fractional nonlinear grey Bernoulli model with grey wolf optimizer. Environ Sci Pollut Res 28, 38128-38144. https://doi.org/10.1007/s11356-021-12736-w

Xiong, P., Dang, Y., Yao, T., Wang, Z., 2014. Optimal modeling and forecasting of the energy consumption and production in China. Energy 77, 623-634.

Xu, G., Schwarz, P., Yang, H., 2019. Determining China's CO2 emissions peak with a dynamic nonlinear artificial neural network approach and scenario analysis. Energy Policy 128, 752-762. https://doi.org/10.1016/j.enpol.2019.01.058

Yu, S., Zheng, S., Li, X., Li, L., 2018. China can peak its energy-related carbon emissions before 2025: Evidence from industry restructuring. Energy Economics 73, 91-107. https://doi.org/10.1016/j.eneco.2018.05.012

Yuan, J., Xu, Y., Hu, Z., Zhao, C., Xiong, M., Guo, J., 2014. Peak energy consumption and CO2 emissions in China. Energy Policy 68, 508-523. https://doi.org/10.1016/j.enpol.2014.01.019

Zapata, H.O., Rambaldi, A.N., 1997. Monte Carlo evidence on cointegration and causation. Oxford Bulletin of Economics and statistics 59, 285-298.

Zeeshan, M., han, J., Rehman, A., Ullah, I., Hussain, A., Alam Afridi, F.E., 2022. Exploring symmetric and asymmetric nexus between corruption, political instability, natural resources and economic growth in the context of Pakistan. Resources Policy 78, 102785. https://doi.org/10.1016/j.resourpol.2022.102785

Zhang, C., Nian, J., 2013. Panel estimation for transport sector CO2 emissions and its affecting factors: A regional analysis in China. Energy Policy 63, 918-926.

Zhang, C., Tian, H., Pan, S., Lockaby, G., Chappelka, A., 2014. Multi-factor controls on terrestrial carbon dynamics in urbanized areas. Biogeosciences 11, 7107-7124. https://doi.org/10.5194/bg-11-7107-2014

Zhang, L., Long, R., Chen, H., Geng, J., 2019. A review of China's road traffic carbon emissions. Journal of

Cleaner Production 207, 569-581. https://doi.org/10.1016/j.jclepro.2018.10.003

Zhang, Q., Li, Z., Wang, G., Li, H., 2016. Study on the impacts of natural gas supply cost on gas flow and infrastructure deployment in China. Applied Energy 162, 1385-1398.

Zhang, S., Hu, T., Li, J., Cheng, C., Song, M., Xu, B., Baležentis, T., 2019. The effects of energy price, technology, and disaster shocks on China's Energy-Environment-Economy system. Journal of Cleaner Production 207, 204-213. https://doi.org/10.1016/j.jclepro.2018.09.256

Zhang, X.-P., Cheng, X.-M., 2009. Energy consumption, carbon emissions, and economic growth in China. Ecological Economics 68, 2706-2712. https://doi.org/10.1016/j.ecolecon.2009.05.011

Zhang, Y.-J., Sun, Y.-F., Huang, J., 2018. Energy efficiency, carbon emission performance, and technology gaps: Evidence from CDM project investment. Energy Policy 115, 119-130. https://doi.org/10.1016/j.enpol.2017.12.056

Zhao, H., Huang, G., Yan, N., 2018. Forecasting energy-related CO2 emissions employing a novel SSA-LSSVM model: considering structural factors in China. Energies 11, 781.

Zhao, X., Du, D., 2015. Forecasting carbon dioxide emissions. Journal of Environmental Management 160, 39-44. https://doi.org/10.1016/j.jenvman.2015.06.002

SCIENTIFIC ARTICLES BASED ON THIS THESIS

APPENDICES

A. 1 LCO VARIABLE STABILITY$_2$

A.1.1 Augmented Dickey-Fuller (ADF) test

Level test		t-Statistic	Prob.*
Augmented Dickey-Fuller test statistic		-3.384909	0.0386
Test critical values:	1% level	-4.297073	
	5% level	-3.212696	
	10% level	-2.747676	

A.1.2 Phillips-Perron test

Level test		Adj. t-Stat	Prob.*
Phillips-Perron test statistic		-4.500640	0.0063
Test critical values:	1% level	-4.200056	
	5% level	-3.175352	
	10% level	-2.728985	

A.2 STATIONARITY OF THE LCSU VARIABLE

A.2. 1Test Augmented Dickey-Fuller

Level test		t-Statistic	Prob.*
Augmented Dickey-Fuller test statistic		1.628491	0.9649
Test critical values:	1% level	-2.792154	
	5% level	-1.977738	
	10% level	-1.602074	

First difference test		t-Statistic	Prob.*
Augmented Dickey-Fuller test statistic		-1.918898	0.0476
Test critical values:	1% level	-2.816740	
	5% level	-1.882344	
	10% level	-1.601144	

A.2.2 Phillips-Perron test

Level test		Adj. t-Stat	Prob.*
Phillips-Perron test statistic		1.408526	0.9492
Test critical values:	1% level	-2.792154	
	5% level	-1.977738	
	10% level	-1.602074	

First difference test		Adj. t-Stat	Prob.*
Phillips-Perron test statistic		-1.918898	0.0476
Test critical values:	1% level	-2.816740	
	5% level	-1.882344	
	10% level	-1.601144	

*MacKinnon (1996) one-sided p-values.

A.3 STATIONARITY TEST FOR THE LCGA VARIABLE

A.3.1 Augmented Dickey-Fuller   test

Level test		t-Statistic	Prob.*
Augmented Dickey-Fuller test statistic		-3.118058	0.0440
Test critical values:	1% level	-4.200056	
	5% level	-3.075352	
	10% level	-2.728985	

A.3.2   Phillips-Perron test

Level test		Adj. t-Stat	Prob.*
Phillips-Perron test statistic		-3.118058	0.0440
Test critical values:	1% level	-4.200056	
	5% level	-3.075352	
	10% level	-2.728985	

A.4 STATIONARITY   OF THE LPSU VARIABLE

A.4. 1 Test Augmented Dickey-Fuller

Level test		t-Statistic	Prob.*
Augmented Dickey-Fuller test statistic		-0.420176	0.5077
Test critical values:	1% level	-2.792154	
	5% level	-1.977738	
	10% level	-1.602074	

First difference test		t-Statistic	Prob.*
Augmented Dickey-Fuller test statistic		-4.360205	0.0006
Test critical values:	1% level	-2.816740	
	5% level	-1.982344	
	10% level	-1.601144	

*MacKinnon (1996) one-sided p-values.

A.4.2 Phillips-Perron   test

Level test		Adj. t-Stat	Prob.*
Phillips-Perron test statistic		-0.680317	0.3990
Test critical values:	1% level	-2.792154	
	5% level	-1.977738	
	10% level	-1.602074	

First difference test		Adj. t-Stat	Prob.*
Phillips-Perron test statistic		-4.498845	0.0004
Test critical values:	1% level	-2.816740	
	5% level	-1.982344	
	10% level	-1.601144	

A.5 STATIONARITY   OF THE LPGA VARIABLE

A.5.1 Augmented Dickey-Fuller   test

Level test		t-Statistic	Prob.*
Augmented Dickey-Fuller test statistic		-0.943847	0.2844
Test critical values:	1% level	-2.816740	
	5% level	-1.982344	
	10% level	-1.601144	

First difference test		t-Statistic	Prob.*
Augmented Dickey-Fuller test statistic		-4.674892	0.0003
Test critical values:	1% level	-2.816740	
	5% level	-1.982344	
	10% level	-1.601144	

A.5.2   Phillips-Perron test

Level test		Adj. t-Stat	Prob.*
Phillips-Perron test statistic		-0.650502	0.4123
Test critical values:	1% level	-2.792154	
	5% level	-1.977738	
	10% level	-1.602074	

First difference test		Adj. t-Stat	Prob.*
Phillips-Perron test statistic		-4.640808	0.0004
Test critical values:	1% level	-2.816740	
	5% level	-1.982344	
	10% level	-1.601144	

A.6 STATIONARITY   OF THE LPAU VARIABLE

A.6. 1Test Augmented Dickey-Fuller

Level test		t-Statistic	Prob.*
Augmented Dickey-Fuller test statistic		-0.763134	0.3625
Test critical values:	1% level	-2.792154	
	5% level	-1.977738	
	10% level	-1.602074	

First difference test		t-Statistic	Prob.*
Augmented Dickey-Fuller test statistic		-3.085982	0.0059
Test critical values:	1% level	-2.816740	
	5% level	-1.982344	
	10% level	-1.601144	

A.6.2 Phillips-Perron test

Level test		Adj. t-Stat	Prob.*

Phillips-Perron test statistic		-1.442763	0.1130
Test critical values:	1% level	-2.792154	
	5% level	-1.977738	
	10% level	-1.602074	

First difference test		Adj. t-Stat	Prob.*
Phillips-Perron test statistic		-4.311583	0.0006
Test critical values:	1% level	-2.816740	
	5% level	-1.982344	
	10% level	-1.601144	

A.7 STATIONARITY OF THE LURB VARIABLE

A.7.1 Test Augmented Dickey-Fuller

Level test		t-Statistic	Prob.*
Augmented Dickey-Fuller test statistic		-3.537556	0.0308
Test critical values:	1% level	-4.297073	
	5% level	-3.212696	
	10% level	-2.747676	

A.7.2 Phillips-Perron test

Level test		Adj. t-Stat	Prob.*
Phillips-Perron test statistic		1.798122	0.9738
Test critical values:	1% level	-2.792154	
	5% level	-1.977738	
	10% level	-1.602074	

First difference test		Adj. t-Stat	Prob.*

Phillips-Perron test statistic		-1.968411	0.0487
Test critical values:	1% level	-2.816740	
	5% level	-1.782344	
	10% level	-1.601144	

A.8 ESTIMATION OF   THE ARDL MODEL

Model 1

Variable	Coefficient	Std. Error	t-Statistic	Prob.*
LCO2(-1)	1.530343	0.250280	6.114533	0.0088
LCSU	-0.044877	0.183946	-0.243968	0.8230
LCSU(-1)	-0.502506	0.094683	-5.307245	0.0131
LPSU	1.326800	0.241682	5.489869	0.0119
LPSU(-1)	-0.908701	0.257737	-3.525690	0.0388
LPAU	0.116668	0.083449	1.398072	0.2565
LPAU(-1)	0.077904	0.044183	1.763240	0.1761
LURB	1.414361	2.295467	0.616154	0.5814

R-squared	0.996824	Mean dependent var	7.482181
Adjusted R-squared	0.989413	S.D. dependent var	0.372969
S.E. of regression	0.038377	Akaike info criterion	-3.527468
Sum squared resid	0.004418	Schwarz criterion	-3.238090

Log likelihood	27.40107	Hannan-Quinn criter.	-3.709881
Durbin-Watson stat	2.907866		

Model 2

Variable	Coefficient	Std. Error	t-Statistic	Prob.*
LCO2(-1)	1.196371	0.604495	1.979125	0.1864
LCGA	0.102556	0.472682	0.216966	0.8484
LCGA(-1)	-0.363783	0.237313	-1.532924	0.2650
LPGA	2.083778	0.620590	3.357734	0.0784
LPGA(-1)	-1.178990	0.330604	-3.566165	0.0704
LPAU	0.245617	0.151936	1.616588	0.2474
LPAU(-1)	-0.145649	0.152750	-0.953513	0.4410
LURB	5.415729	2.177097	2.487592	0.1307
LURB(-1)	-7.715484	1.740924	-4.431833	0.0473

R-squared	0.996872	Mean dependent var	7.482181

Adjusted R-squared	0.984359	S.D. dependent var	0.372969
S.E. of regression	0.046644	Akaike info criterion	-3.360913
Sum squared resid	0.004351	Schwarz criterion	-3.035362
Log likelihood	27.48502	Hannan-Quinn criter.	-3.566127
Durbin-Watson stat	2.107542		

A.9 ARDL DIAGNOSTIC TEST RESULTS AND MODEL STABILITY

Model 1

Jarque-Bera (Normality test)

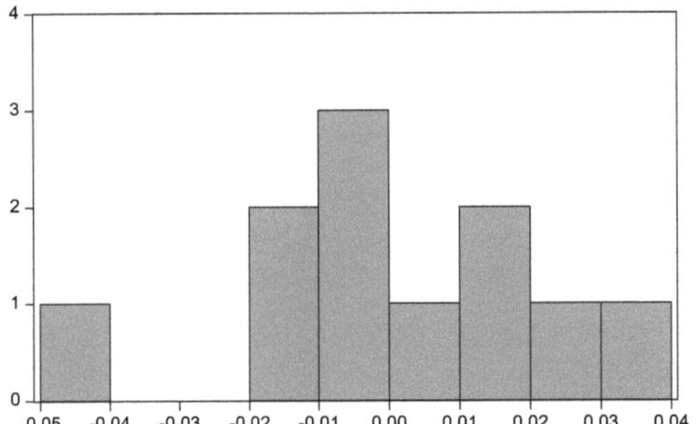

Breusch-Godfrey Serial Correlation LM Test:

F-statistic	0.789029	Prob. F(1,2)	0.4681
Obs*R-squared	3.111951	Prob. Chi-Square(1)	0.0777

Heteroskedasticity Test: Breusch-Pagan-Godfrey

F-statistic	0.285500	Prob. F(8,2)	0.9192
Obs*R-squared	5.864611	Prob. Chi-Square(8)	0.6624
Scaled explained SS	0.366245	Prob. Chi-Square(8)	1.0000

Heteroskedasticity Test: ARCH

F-statistic	0.659293	Prob. F(1,8)	0.4403
Obs*R-squared	0.761370	Prob. Chi-Square(1)	0.3829

Ramsey RESET Test

	Value	df	Probability
t-statistic	0.485935	2	0.6750
F-statistic	0.236133	(1, 2)	0.6750

Model 2

Jarque-Bera (Normality test)

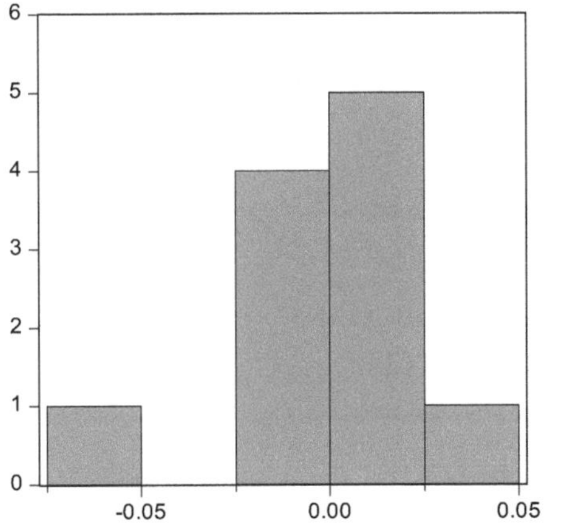

Breusch-Godfrey Serial Correlation LM Test:

F-statistic	0.036703	Prob. F(1,1)	0.8795
Obs*R-squared	0.389434	Prob. Chi-Square(1)	0.5326

Heteroskedasticity Test: Breusch-Pagan-Godfrey

F-statistic	2.071484	Prob. F(9,1)	0.4953
Obs*R-squared	10.44001	Prob. Chi-Square(9)	0.3161
Scaled explained SS	0.713099	Prob. Chi-Square(9)	0.9999

Heteroskedasticity Test: ARCH

F-statistic	0.081903	Prob. F(1,8)	0.7820
Obs*R-squared	0.101342	Prob. Chi-Square(1)	0.7502

Ramsey RESET Test

	Value	df	Probability
t-statistic	11.05936	1	0.0574
F-statistic	122.3094	(1, 1)	0.0574

A.10 RESULTS OF THE CO-INTEGRATION TEST

Model 1

Test Statistic	Value	Meaning.	I(0)	I(1)
			Asymptotic: n=1000	
F-statistic	27.39011	10%	1.9	3.01
k	4	5%	2.26	3.48
		2.5%	2.62	3.9
		1%	3.07	4.44

Model 2

Test Statistic	Value	Meaning.	I(0)	I(1)
			Asymptotic: n=1000	
F-statistic	31.33710	10%	1.9	3.01
k	4	5%	2.26	3.48
		2.5%	2.62	3.9
		1%	3.07	4.44

A.   11LONG-TERM AND SHORT-TERM DYNAMICS

Model 1

Levels Equation

Case 1: No Constant and No Trend

Variable	Coefficient	Std. Error	t-Statistic	Prob.
LCSU	1.032130	0.255539	4.039035	0.0273
LPSU	-0.788356	0.409104	-1.927031	0.1496
LPAU	0.366880	0.123391	-2.973317	0.0589
LURB	-2.666882	3.881973	-0.686991	0.5414

ECM Regression

Case 1: No Constant and No Trend

Variable	Coefficient	Std. Error	t-Statistic	Prob.
D(LCSU)	-0.044877	0.035809	-1.253213	0.2989

Variable	Coefficient	Std. Error	t-Statistic	Prob.
D(LPSU)	1.32680	0.071525	18.55006	0.00030
D(LPAU)	0.11666	0.026959	4.327613	0.02278
CointEq(-1)*	-0.530343	0.029668	17.87600	0.0004

Model 2

Levels Equation

Case 1: No Constant and No Trend

Variable	Coefficient	Std. Error	t-Statistic	Prob.
LCGA	1.33027	1.235509	1.076705	0.39429
LPGA	-4.607556	9.953909	-0.462889	0.6889
LPAU	-0.509080	0.568267	-0.895847	0.4649
LURB	11.7113	37.89655	0.309034	0.78650

ECM Regression

Case 1: No Constant and No Trend

Variable	Coefficient	Std. Error	t-Statistic	Prob.
D(LCGA)	0.10255	0.037594	2.727982	0.11226
D(LPGA)	2.08377	0.082962	25.11736	0.00168
D(LPAU)	0.24561	0.028566	8.598376	0.01337
D(LURB)	5.41572	0.556560	9.730721	0.01049
CointEq(-1)*	-0.196371	0.009057	21.68079	0.0021

A.12 TODA-YAMAMOTO CAUSALITY TEST

Model 1

Dependent variable: LCO2

Excluded	Chi-sq	df	Prob.
LCSU	5.710915	1	0.0091
LPSU	9.891653	1	0.0017
LPAU	0.754317	1	0.3851
LURB	0.229185	1	0.6321
All	11.23133	4	0.0241

Dependent variable: LCSU

Excluded	Chi-sq	df	Prob.
LCO2	0.021974	1	0.8822
LPSU	28.30917	1	0.0000
LPAU	0.637192	1	0.4247
LURB	0.023391	1	0.8784
All	35.02920	4	0.0000

Dependent variable: LPSU

Excluded	Chi-sq	df	Prob.
LCO2	11.44935	1	0.0007
LCSU	3.634907	1	0.0566
LPAU	2.090497	1	0.1482
LURB	0.004129	1	0.9488
All	22.36893	4	0.0002

Dependent variable: LPAU

Excluded	Chi-sq	df	Prob.
LCO2	3.167163	1	0.0751
LCSU	5.140884	1	0.0234
LPSU	0.223278	1	0.6366
LURB	6.070668	1	0.0137
All	6.195467	4	0.1850

Dependent variable: LURB

Excluded	Chi-sq	df	Prob.
LCO2	12.48507	1	0.0004
LCSU	2.114774	1	0.1459
LPSU	10.75021	1	0.0010
LPAU	0.042611	1	0.8365

	70.70852	4	0.0000
All	70.70852	4	0.0000

Model 2

Dependent variable: LCO2

Excluded	Chi-sq	df	Prob.
LCGA	6.178338	1	0.0028
LPGA	8.720829	1	0.0031
LPAU	3.219594	1	0.0728
LURB	0.090894	1	0.7630
All	10.23227	4	0.0367

Dependent variable: LCGA

Excluded	Chi-sq	df	Prob.
LCO2	21.90388	1	0.0000
LPGA	1.754657	1	0.1853
LPAU	9.034819	1	0.0026
LURB	1.201672	1	0.2730
All	41.33283	4	0.0000

Dependent variable: LPGA

Excluded	Chi-sq	df	Prob.
LCO2	45.82942	1	0.0000
LCGA	9.454016	1	0.0021
LPAU	1.891916	1	0.1690
LURB	13.45406	1	0.0002
All	51.14815	4	0.0000

Dependent variable: LPAU

Excluded	Chi-sq	df	Prob.
LCO2	2.268332	1	0.1320
LCGA	2.487601	1	0.1147
LPGA	0.348337	1	0.5551
LURB	0.930078	1	0.3348
All	3.191431	4	0.5263

Dependent variable: LURB

Excluded	Chi-sq	df	Prob.
LCO2	21.45237	1	0.0000
LCGA	3.054399	1	0.0805

LPGA	13.22336	1	0.0003
LPAU	5.151011	1	0.0232
All	69.55314	4	0.0000

yes

I want morebooks!

Buy your books fast and straightforward online - at one of world's fastest growing online book stores! Environmentally sound due to Print-on-Demand technologies.

Buy your books online at
www.morebooks.shop

Kaufen Sie Ihre Bücher schnell und unkompliziert online – auf einer der am schnellsten wachsenden Buchhandelsplattformen weltweit! Dank Print-On-Demand umwelt- und ressourcenschonend produziert.

Bücher schneller online kaufen
www.morebooks.shop

Printed by Books on Demand GmbH, Norderstedt / Germany